UNITS, SYMBOLS, AND TERMINOLOGY FOR PLANT PHYSIOLOGY

UNITS, SYMBOLS, AND TERMINOLOGY FOR PLANT PHYSIOLOGY

A Reference for Presentation of Research Results in the Plant Sciences

Sponsored by the International Association for Plant Physiology

Frank B. Salisbury, Editor
Utah State University

New York Oxford
Oxford University Press
1996

Oxford University Press

Oxford New York
Athens Auckland Bangkok Bogota Bombay
Buenos Aires Calcutta Cape Town Dar es Salaam
Delhi Florence Hong Kong Istanbul Karachi
Kuala Lumpur Madras Madrid Melbourne
Mexico City Nairobi Paris Singapore
Taipei Tokyo Toronto

and associated companies in
Berlin Ibadan

Published by Oxford University Press, Inc.
198 Madison Avenue, New York, New York 10016

Library of Congress Cataloging-in-Publication Data
Units, symbols, and terminology for plant physiology : a reference for presentation of research results in the plant sciences / sponsored by the International Association for Plant Physiology ; Frank B. Salisbury, editor.
p. cm.
Includes bibliographical references and index.
ISBN 0-19-509445-X
1. Plant physiology—Terminology. 2. Botany—Terminology.
3. Technical writing. I. Salisbury, Frank B.
II. International Association for Plant Physiology.
QK710.5.U55 1996
581.1'014—dc20 95-50593

9 8 7 6 5 4 3 2 1
Printed in the United States of America
on acid-free paper

CONTENTS

CONTRIBUTORS

Mary Barkworth, Utah State University, U.S.A.
Clanton C. Black, University of Georgia, U.S.A.
Bruce G. Bugbee, Utah State University, U.S.A.
Robert E. Cleland, University of Washington, U.S.A.
Jack Dainty, University of Toronto, Canada
Ralph O. Erickson, University of Pennsylvania, U.S.A.
Leslie H. Fuchigami, Oregon State University, U.S.A.
Donald R. Geiger, University of Dayton, Ohio, U.S.A.
Wolfgang W. Haupt, Universtät Erlangen-Nürnberg, Germany
Ross E. Koning, Eastern Connecticut University, U.S.A.
Willard L. Koukkari, University of Minnesota, U.S.A.
Donald T. Krizek, U.S. Department of Agriculture, Beltsville, Maryland, U.S.A.
James M. Lyons, University of California, Davis, California, U.S.A.
Eugene V. Maas, U.S. Salinity Laboratory, USDA-ARS, Riverside, California, U.S.A.
John McNeill, Royal Ontario Museum, Toronto, Canada
Carl A. Price, Rutgers University, New Jersey, U.S.A.
William Rains, University of California, Davis, California, U.S.A.
John K. Raison, Macquarie University, North Ryde, Australia (deceased)
Ellen Reardon, Rutgers University, New Jersey, U.S.A.
John C. Sager, Kennedy Space Center, Florida, U.S.A.
Frank B. Salisbury, Utah State University, U.S.A.
Michael J. Savage, University of Natal, Pietermaritzburg, Republic of South Africa.
Kenneth Shackel, University of California, Davis, California, U.S.A.
Donald Sisson, Utah State University, U.S.A.
Beatrice M. Sweeney, University of California at Santa Barbara, U.S.A. (deceased)
Theodore W. Tibbitts, University of Wisconsin-Madison, U.S.A.
Aart J.E. van Bel, Justus-Liebig Universität, Giessen, Germany

PREFACE

When one person wishes to communicate some information directly to another person, it is essential that the two speak the same language; that is, the words and symbols must have the same meaning for both persons. Such a thought provides one motivation for the preparation of this book, which is designed to be a reference source for plant physiologists and other plant scientists who are preparing their research results for publication or other presentation. The primary goal is to provide information about the use of units, symbols, and terminology in the plant sciences, especially plant physiology. In addition, we also provide some hints and instructions about writing and the preparation of posters and slide presentations for scientific meetings, including a format for presentation of growth-chamber data.

Section I introduces the basics. Its three chapters consider the use of SI units, rules for botanical nomenclature, and basic principles of statistics. Sections II, III, and IV present more detail in the fields of plant biophysics, biochemistry, and growth and development. These sections emphasize SI units whenever that is appropriate, but they also contain many lists of terms that are used in the plant sciences. The appendices contain the hints and instructions for writing and for preparing posters and slide presentations, plus a summary of guidelines for reporting environmental parameters for plant experiments in controlled environments. The chapter on biochemistry was modified from *The Journal of Biological Chemistry;* it is included here as a handy reference. Appendix C was also prepared for another publication. All other sections were originally prepared for this volume.

Each chapter was first prepared by one or more specialists in the field, and the authors then sent their chapters to several colleagues. As a result, the present chapters represent at least the beginnings of a consensus about the terms and sometimes symbols within each subfield. Although the time when all plant scientists agree on all units, symbols, terminology, and presentation techniques may be in the distant future (if it ever arrives), it is hoped that this book will bring us closer to such a meeting of the minds. After I had edited the manuscripts sent by the various authors, the entire book was sent to each author, who often commented about some chapters besides his or her own. This process was repeated several times over a period exceeding a decade (mostly because the project was set aside several times while other projects were being completed). During this long gestation period, two authors died and several others retired! In spite of the long period from conception to birth, every chapter includes significant changes made shortly before publication. The book presents the most current thinking of its authors and editor.

The chapters that include definitions of terms follow two different approaches: In some chapters, terms are arranged alphabetically; in others they follow an order in which one term builds upon the preceding term or terms (a mini-review of the subject). The choice of approach depended upon the author and the subject matter. In the non-alphabetical cases, the number of terms is rather limited so that it should be relatively easy to find a term by scanning the lists.

A few references are presented, especially where definitions are somewhat controversial. And controversy remains in plant physiology! Please submit suggestions for future editions to me or to the appropriate chapter author.

We have tried to remove inconsistencies and contradictions although some seem to be inevitable. We are aware of considerable redundancy, which should make the booklet easier to use as a reference source. An editorial inconsistency that I have decided to allow concerns the use of references. Many show only initials of authors, but when given names were known to me, I included them. We have followed a reference style that includes written-out journal names rather than abbreviations and more punctuation than is used in many current journals. This takes a little more space, but we believe it will make it easier for a reader to use the references.

Several secretaries were involved with the manuscript, but Laura Wheelwright did much formatting, and Mary Ann Clark must have spent the equivalent of an intense, full-time year working on the final formatting of camera-ready copy with much direction from Kirk Jensen, a Senior Editor at the Oxford University Press. The authors and I wish to express much appreciation to those diligent secretaries; their efforts were often "above and beyond the call of duty."

F.B. Salisbury
Logan, Utah

APPROXIMATE CONVERSIONS: METRIC ↔ U.S.[a]

Temperature		Length	Mass	Volume
$C = (5/9)(F - 32)$	$F = (9C/5) + 32$	1 cm = 0.3937 in	1 kg = 2.205 lb	1 L = 1.0567 qt
		1 in = 2.54 cm	1 g = 0.0353 oz	100 mL = 3.38 oz
		1 m = 1.0936 yd	1 lb = 0.4536 kg	1 oz = 29.57 mL
		1 m = 3.281 ft	1 oz = 28.35 g	1 qt = 0.946 L
		1 yd = 0.914 m		1 gal = 3.785 L
		1 ft = 0.305 m		

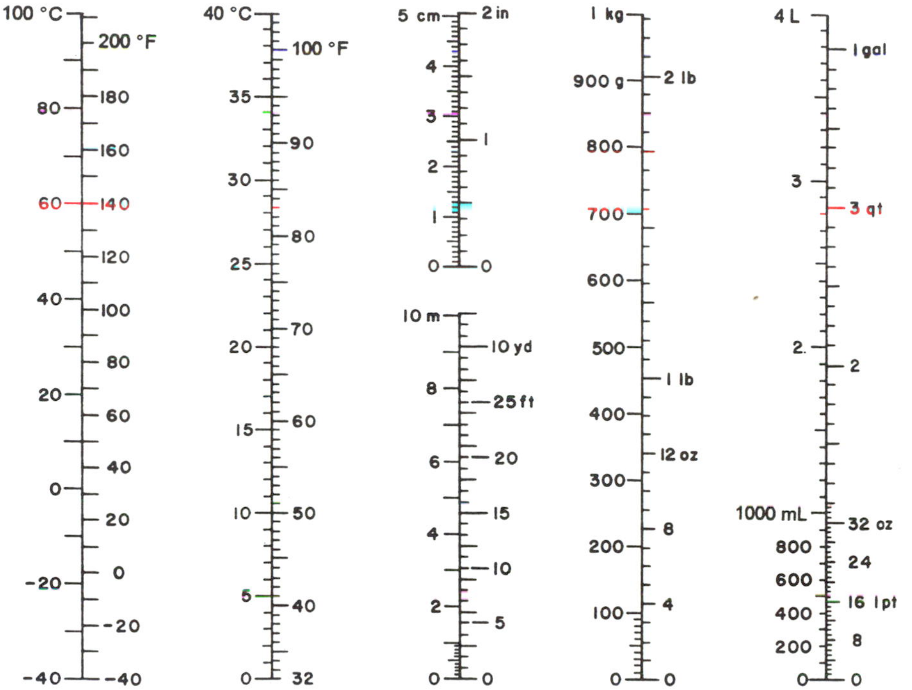

[a] This chart was prepared by F.B.S. for: Frank B. Salisbury and Cleon W. Ross. 1969. Plant Physiology, First Edition. Wadsworth Publishing Co., Inc., Belmont, California. It was not used in subsequent editions. Some letters have been changed to reflect the conventions presented in this book.

GREEK ALPHABET AND ENGLISH EQUIVALENTS

Greek letter (roman)	Greek letter (*italic*)	Greek name	English equivalent (phonetic)
Α α	*Α α*	Alpha	ä
Β β	*Β β*	Beta	b
Γ γ	*Γ γ*	Gamma	g
Δ δ	*Δ δ*	Delta	d
Ε ϵ	*Ε ϵ*	Epsilon	e
Ζ ζ	*Ζ ζ*	Zeta	z
Η η	*Η η*	Eta	e (ē)
Θ θ	*Θ θ*	Theta	th
Ι ι	*Ι ι*	Iota	i
Κ κ	*Κ κ*	Kappa	k
Λ λ	*Λ λ*	Lambda	l
Μ μ	*Μ μ*	Mu	m
Ν ν	*Ν ν*	Nu	n
Ξ ξ	*Ξ ξ*	Xi	ks, x
Ο ο	*Ο ο*	Omicron	o
Π π	*Π π*	Pi	p
Ρ ρ	*Ρ ρ*	Rho	r
Σ σ ς[a]	*Σ σ ς*[a]	Sigma	s
Τ τ	*Τ τ*	Tau	t
Υ υ	*Υ υ*	Upsilon	y
Φ φ	*Φ φ*	Phi	f, ph
Χ χ	*Χ χ*	Chi	ch, kh
Ψ ψ	*Ψ ψ*	Psi	ps
Ω ω	*Ω ω*	Omega	o (ō)

[a] At end of word.

UNITS, SYMBOLS, AND TERMINOLOGY FOR PLANT PHYSIOLOGY

I

THE BASICS

This section deals mostly with constructed scientific languages. How do people who want to communicate usually achieve a common language? Mostly, we begin as infants and just *use* the language until meanings become clear. But there are problems with this approach. For one thing, people in different parts of society—different geographical areas, for example—have formed different languages. Furthermore, usage often produces languages that lack logic and consistency. As scientists, we would like to communicate effectively with everyone else on the planet who might share our common interests. One solution that seems to be falling into place without any directed effort is the broadening acceptance of English as the language of science (and much of commerce, etc.). A second solution for science has involved a conscious and directed mental effort to *create* consistency and uniformity. Groups of scientists have tried to find ways to *agree* on how to express physical quantities, nomenclature of organisms, and mathematical symbols (among other things). In this section, we present the three constructed languages that deal with physical quantities, taxonomic nomenclature, and statistics:

1. The International System of Units for expressing physical quantities,

2. The adopted conventions for naming plant material; that is, many of the important rules of taxonomic nomenclature agreed upon in Botanical Congresses, and

3. Statistical procedures and their notations; these provide a measure of significance.

All plant scientists who work with quantitative measurements, regardless of their specialty within the field of plant physiology or in other areas of botany, need to be conversant with these two international systems of communication plus the means of evaluating the reliability of their numerical data.

1

THE INTERNATIONAL SYSTEM OF UNITS (SI UNITS)[1]

Frank B. Salisbury
Plants, Soils, and Biometeorology Department
Utah State University
Logan, Utah 84322-4820, U.S.A.

As modern science came into being, it depended more and more upon the accurate measurement of physical quantities. Such measurement requires a system of standards that is recognized and accepted by all those who would communicate their measurements to each other. In response to this need, the **metric system** of measurement was devised during the French Revolution (1789 to 1799). It was an attempt to devise a decimal system of measures that would simplify and unify calculations. Nearly a century later, recognizing the need to further improve the system, the *Bureau International des Poids et Mesures* (BIPM) was set up by the *Convention du Mètre* signed in Paris in 1875 by seventeen States; the *Convention* was amended in 1921. The task of the BIPM is to ensure worldwide unification of physical measurements. It operates in offices and laboratories in Sèvres, near Paris, France, under supervision of the *Comité International des Poids et Mesures* (CIPM), which consists of 18 members, each from a different State. The CIPM itself comes under the authority of the *Conférence Générale des Poids et Mesures* (CGPM), which consists of delegates from all the Member States (46 States in March, 1991) of the *Convention du Mètre*. The CGPM meets at present every four years, but the CIPM meets every year.

By the mid twentieth century, the metric system was being widely used in science, but in many cases, individual branches of science had developed their own specialized units and terms. For example, the CGS (centimeter·gram·second) system of mechanical units, used especially in physics, included such terms as dyne, erg, poise, stokes, gauss, oersted, and maxwell (all now considered obsolete). To

---

[1] Early versions of this chapter were published as Appendix A in *Plant Physiology, Fourth Edition*, by F. B. Salisbury and Cleon W. Ross, published by Wadsworth Publishing Company, Belmont, California, 94002, U.S.A., and in **Journal of Plant Physiology** (Salisbury, 1991). Recent study of the first and second-level authorities (described in this chapter) has led not only to a somewhat different approach but also to some important modifications and changes in a few units and the rules for their use.

unify the metric system further, the 9th CGPM in 1948 instructed the CIPM to study and recommend the establishment of a "*practical system of units of measurement* suitable for adoption by all signatories to the Metre Convention" (see Taylor, 1991). This conference also laid down a set of principles for unit symbols and gave a list of units with special names. Six base units were established in 1954, and the 11th CGPM in 1960 adopted the name ***Le Système International d'Unités*** (English: **International System of Units**) with the international abbreviation **SI**. The 11th CGPM also laid down rules for prefixes, derived and supplementary units, and other matters. Thus the SI was born in 1960, and subsequent meetings have added various amendments. The 14th CGPM, in 1971, for example, added the mole to the original six base units, making a total of seven base units in the SI, each with its own name and symbol, which is the same (with slight spelling differences) in all languages. The SI is currently by far the best measurement system humankind has been able to develop.

The purpose of this chapter is to present the SI, especially as it applies to the plant sciences. The information presented here comes from various sources. It is convienent to think of three levels of authority: The *first*, most primary level, is "*Le Système International d'Unités (SI), 6e Édition*", French and English texts. This is the definitive publication issued in 1991 by the International Bureau of Weights and Measures (BIPM). Although this publication was prepared jointly with the National Physical Laboratory in the United Kingdom, some words and practices follow United States rather than British usage. In general, this usage (e.g., meter instead of metre and liter instead of litre) is closer to the European usage than are the British practices. The United States translation of this primary volume is listed in the references to this chapter as Taylor (1991); it is virtually identical to the version published by the BIPM except for a few small matters such as use of the dot instead of the comma as the decimal marker. One *second* level source of authority is the ISO Standards Handbook, Third Edition, published in 1993 by the International Organization for Standardization (ISO) in Geneva, Switzerland. It expands upon the rules of the primary source, and these expansions have influenced the deliberations of the CGPM so that some recommendations of the ISO Standards Handbook have become official SI rules. Another *second* level source of authority is Special Publication (SP) 811 of the National Institute of Standards and Technology (NIST, formerly the U.S. National Bureau of Standards). SP 811 is the "Guide for the use of the International System of Units (SI)," prepared by Taylor (1995). The *third* level of authority includes the many publications such as this one that attempt to summarize, interpret, and condense the SI for a given field. (See many of the **bold-faced** entries in the list at the end of this chapter.) The rules presented here are taken almost exclusively from the first two authority levels.

1. QUANTITIES AND UNITS

In science in general and the plant sciences in particular, we deal with *physical quantities*. To communicate these physical quantities, we use three kinds of symbols: a symbol for the **physical quantity**, a symbol for a **numerical value** (i.e., a number), and a symbol for a **unit**. For example, if we want to communicate the length of some object, we can write:

$$l = 5.67 \text{ m}$$

If this notation is to be a meaningful form of communication, those of us who want to communicate must agree that the symbol for length is l, that we will use Arabic numerals, and that the meter (m) represents a *standard unit* of length; namely, the length of the path travelled by light in vacuum during a time interval of 1/299 792 458 of a second. Of course, for practical purposes, most of us will trust the manufacturers of meter sticks and other measuring devices, assuming that those manufacturers accurately follow a reliable standard when they create the measuring instruments. The various nations have bureaus to insure this accuracy (in the United States, The National Institute of Standards and Technology), and as noted above, the ultimate authority for standards goes back to the CGPM and the BIPM.

Remember that the unit represents a number. The physical quantity is the numerical value multiplied by the unit. Thus, the unit is subject to algebraic manipulations. For example, the numerical value can be thought of as the ratio of the physical quantity to the unit; in the above example: $l/\text{m} = 5.67$. This notation is particularly useful in graphs and in the headings of columns in tables.

Note that the symbol for the physical quantity (length in the example) is written in *italic* or *slanted* type (underlined if italic type is not available), and the symbol for the unit is written in roman (upright) type. This rule (listed again in Table 4, #17) should be followed with Greek symbols as well as those from the Roman alphabet.

2. *LE SYSTÈME INTERNATIONAL D'UNITÉS (SI)*

The SI is a so-called *coherent unit system*, in which the equations between numerical values have exactly the same form (including the numerical factors) as the corresponding equations between the quantities. This is achieved by defining units for the base quantities (the *base units*), and then deriving further units from these base units based upon the equations between the quantitites. For example, the equation for speed (v) shows speed as being equal to the incremental change in distance ($\mathrm{d}l$) divided by the incremental change in time ($\mathrm{d}t$): $v = \mathrm{d}l/\mathrm{d}t$. Thus the unit for speed is the meter per second: m/s. Therefore, the SI includes *base units* and *derived units*. In addition there are two *supplementary units*, the radian (rad, for plane angle) and the steradian (sr, for solid angle).

The seven base units of the SI are the **meter** (length; **metre** is also used, especially in Britain and France), **kilogram** (mass), **second** (time), **ampere** (electric current), **kelvin** (thermodynamic temperature), **candela** (luminous intensity), and **mole** (amount of substance). These units are shown with their symbols in Table 1. Actually, it is possible by combining the units of space (length, area, and volume) with those of mass, time, and temperature to derive units of any physical quantity.

The basic unit of length is the **meter** (m), which was originally defined as equivalent to the length of a bar preserved in Sèvres, France; in 1960 it was defined as the length equal to 1 650 763.73 wave lengths in vacuum of the radiation corresponding to the transition between the levels $2p_{10}$ and $5d_5$ of the krypton-86 atom.

In 1983, again in response to advancing technology, the meter was redefined as the distance light travels in a vacuum during a time interval of 1/299 792 458 of a second.

Table 1. The Seven Base Units

Quantity	Unit	Symbol
Length (*l*)	meter	m
Mass (not weight) (m^a)	kilogram[a]	kg
Time (*t*)	second	s
Electric current (*I*)	ampere	A
Thermodynamic temperature (*T*)	kelvin	K (not °K)
Luminous intensity (*I*)	candela[b]	cd
Amount of substance (*n*, *Q*)	mole[c]	mol

[a] For historical reasons, the kilogram is the SI base unit rather than the gram. It is a unit of *mass* rather than *weight*. Although *weight* is an acceptable synonym for *mass*, plant scientists should be careful to use *mass* instead of *weight* whenever appropriate—which is most of the time. (Note that the quantity *mass* is symbolized with *italic m*, which is not to be confused with roman m for meter. See ISO Standards Handbook, 1993.)

[b] As a unit of luminous intensity, the candela was traditionally based on the sensitivity of the human eye; we know of no application in plant physiology. The lux (lx) is a measure of illuminance based on the candela ($1\ \text{lx} = 1\ \text{cd}\cdot\text{sr}\cdot\text{m}^{-2}$); it has been widely used in plant science but should be avoided.

[c] The mole should always be used to report the amount of a pure substance, and in such cases the type of substance must be specified. To report the amount of a mixture or of an unknown substance, mass must be used.

For historical reasons, the gram is *not* the SI base unit for mass. The **kilogram** is the only base unit with a prefix. It is equal to the mass of the international prototype of the kilogram, made of platinum-iridium, kept at the BIPM under conditions specified by the first CGPM in 1889. Note that *weight* is technically a measure of the *force* produced by gravity, whereas the kilogram is a unit of *mass*. Mass is a fundamental quantity that does not change with the force of gravity (for example, with location). The weight of objects, on the other hand, is about 1 percent less at the equator than at the poles and is 82 percent less on the moon. Thus it is technically incorrect to use the word *weight* in conjunction with the unit kilogram. The proper unit for weight is the newton. (On earth, the weight of a 10 kg mass is about 98 newtons.) *Although in many technical fields and in everyday use the term "weight" is considered as an acceptable synonym for "mass," plant scientists should use the term "mass" whenever it is appropriate.*

A balance *balances* the mass of an unknown object against a defined mass; hence, a balance measures true mass. All balances depend upon an accelerational force for their function, but the magnitude of the accelerational force does not affect the reading. Unfortunately, the magnitude of accelerational force does affect the measurement of mass on electronic "balances" because they are really scales that measure weight. This is usually not a serious problem because the force of gravity is constant for a given location, and electronic balances and spring scales are calibrated with a standard set of objects of known mass.

All objects with a mass also have a volume and thus displace some air, which has a density of 1.205 kg m^{-3} (standard atmospheric pressure, dry air, 20 °C). A correction for this volume displacement would be necessary in some situations (for example, measuring the mass of a helium balloon!), but most plant tissues have a density similar to that of water (1,000 kg m^{-3}), so the correction is only about 0.1 percent.

Note that a quantity of substance can be expressed either in terms of its mass or the number of particles of which it is composed: "The **mole** is the amount of substance of a system that contains as many elementary entities as there are atoms in 0.012 kilogram of carbon 12. When the mole is used, the elementary entities must be specified and may be atoms, molecules, ions, electrons, other particles, or specified groups of such particles" (Taylor, 1991). Plant physiologists and others include photons among the particles that can be expressed in moles, but note that the einstein (a mole of photons) is not an SI unit and should not be used. Note that 1 mol of a substance contains **Avogadro's number** of particles (now defined as the number of atoms in 0.012 kg of carbon 12 ≈ 6.022045 x 10^{23} particles).

Following various astronomical definitions of the **second** (e.g., 1/86 400 of the mean solar day), the second was defined in 1967 as the duration of 9 192 631 770 periods of the radiation corresponding to the transition between the two hyperfine levels of the ground state of the cesium-133 atom. Although the minute, hour, day, week, month, and year are not officially part of SI, plant physiologists will continue to use them when appropriate.

The **ampere** is defined as that constant current required to produce, in vacuum, a force of 2 x 10^{-7} newtons per meter of length between two parallel conductors of infinite length and 1 meter apart. Because force (the newton) is defined in terms of length, mass, and time (see Table 2), current could also be defined in those terms.

The **kelvin** was defined by the CGPM in 1967 as the fraction 1/273.16 of the thermodynamic temperature of the triple point of water. That same CGPM also adopted the name *kelvin* (symbol K) to be used instead of *degree Kelvin* (symbol °K). In addition to the physical quantity *thermodynamic temperature* (symbol T, unit K), use is also made of Celsius temperature (symbol t, unit °C), where $t = T - T_0$ and $T_0 = 273.15$ K by definition. An interval or difference of Celsius temperature can be expressed in kelvins as well as in degrees Celsius.

Luminous intensity (the **candela**) was defined in terms of the light intensity perceived by the human eye as compared with the intensity of freezing platinum, but in 1979 it was redefined as monochromatic radiation with a frequency of 540 x 10^{12} hertz and a radiant intensity of 1/683 watt per steradian. The watt (unit for power) also combines length, mass, and time. Thus, although the SI recognizes seven base units, only the units of length, mass, time, temperature, and number (the mole) are truly basic in that they are not derived from any other units—and temperature could be derived from the first three.

Because the candela and its derivatives were based on the sensitivity of the human eye, and plant sensitivities may be very different (depending on the pigment involved), the candela and its derivatives (e.g., the lux) should not be used by plant scientists. This is true in spite of the more recent definition based on monochromatic light. While the candela is of value to engineers who are concerned with artificial lighting for human beings, other measures of radiation can be derived from power (the watt) per unit area (W m^{-2}) or the number (moles) of photons per unit area times unit time (usually μmol m^{-2} s^{-1}). These units should be used by plant scientists. In either case, wave lengths or frequencies must be specified.

Table 2 lists the prefixes that are used in the International System of Units. Some third-level publications have suggested that four of the prefixes were "non-preferred": centi, deci, hecto, and deka. Although they were commonly used in the metric system, it was suggested that they should be avoided when it is convenient to use the others. The first-level and second-level sources, however, make no such distinction about being preferred or non-preferred. In many cases, using those prefixes is convenient and leads to clarity. In other cases, it is logical to use only prefixes that differ by a factor of 1 000 (10^3).

Table 2. SI Prefixes[a] (multiples and submultiples)

Prefix	Symbol	Factor	Prefix	Symbol	Factor
deka	da	(10)	deci	d	(10^{-1})
hecto	h	(10^2)	centi	c	(10^{-2})
kilo	k	(10^3)	milli	m	(10^{-3})
mega	M	(10^6)	micro	μ	(10^{-6})
giga	G	(10^9)	nano	n	(10^{-9})
tera	T	(10^{12})	pico	p	(10^{-12})
peta	P	(10^{15})	femto	f	(10^{-15})
exa	E	(10^{18})	atto	a	(10^{-18})
zetta	Z	(10^{21})	zepto	z	(10^{-21})
yotta	Y	(10^{24})	yocto	y	(10^{-24})

[a] The first syllable of every prefix is accented to assure that the prefix will retain its identity.

Table 3 shows some important SI derived units with special names that are derived from the base units and are of value to plant scientists. (See Taylor, 1991, for complete lists.) Note that the standard acceleration due to gravity is an experimentally determined unit, and the unified atomic mass is an arbitrary unit.

Table 3. Derived Units of Interest to Plant Physiologists

Quantity (symbol)[a]	Unit Name	Symbol	Definition
Area (A)	square meter	m^2	m·m
Volume (V)	cubic meter	m^3	m·m·m
Speed or velocity[b] (v)	meters per second	$m \cdot s^{-1}$	$m \cdot s^{-1}$
Force (F)	newton	N	$kg \cdot m \cdot s^{-2}$
Energy (E), work (W), heat (Q)	joule	J	N·m ($m^2 \cdot kg \cdot s^{-2}$)
Power (P)	watt	W	$J \cdot s^{-1}$ ($m^2 \cdot kg \cdot s^{-3}$)
Pressure (p)	pascal	Pa	$N \cdot m^{-2}$ ($kg \cdot s^{-2} \cdot m^{-1}$)
Frequency (ν, Greek nu)	hertz	Hz	cycle s^{-1}
Electric charge (Q)	coulomb	C	A·s
Electric potential (V, φ)	volt	V	$W \cdot A^{-1}$ ($J \cdot A^{-1} \cdot s^{-1}$; $J \cdot C^{-1}$)
Electric resistance (R)	ohm	Ω	$V \cdot A^{-1}$
Electric conductance (G)	siemens	S	$A \cdot V^{-1}$ (Ω^{-1})
Electric capacitance (C)	farad	F	$C \cdot V^{-1}$
Concentration (c)	moles per cubic meter	$mol \cdot m^{-3}$	$mol \cdot m^{-3}$
Irradiance (energy: E)	watts per square meter	$W \cdot m^{-2}$	$J \cdot s^{-1} \cdot m^{-2}$
Irradiance (moles of photons)	moles per square meter second	$mol \cdot m^{-2} \cdot s^{-1}$	$mol \cdot m^{-2} \cdot s^{-1}$
Spectral irradiance (moles of photons)	moles per square meter second nano-meter	$mol \cdot m^{-2} \cdot s^{-1} \cdot nm^{-1}$	$mol \cdot m^{-2} \cdot s^{-1} \cdot nm^{-1}$
Magnetic field strength (H)	amperes per meter	$A \cdot m^{-1}$	$A \cdot m^{-1}$
Activity (of radioactive source: A)	becquerel	Bq	s^{-1}
Standard acceleration due to gravity	standard acceleration due to gravity	g_n	9.806 65 m s^{-2}
Unified atomic mass unit	Unified atomic mass unit	u	1/12 of mass of ^{12}C

[a] Quantity symbols are from ISO Standards Handbook, 1993. There are many variations of such symbols.

[b] Technically, velocity is a vector quantity requiring specification of a magnitude (speed) and a direction, but magnitude is most important in plant science.

Table 4 summarizes the style conventions that govern the use of SI units and that are of interest to plant scientists. Most of these rules are from the primary source of authority: *Le Système International d'Unités (SI), 6e Édition* (or the American English equivalent: Taylor, 1991), but many rely on the second-level ISO Standards Handbook and NIST Special Publication 811 (Taylor, 1995). A few are recommendations from third-level publications; these are noted here and discussed further in the text and in relation to Tables 5 and 6.

Table 4. Summary of SI Style Conventions (Rules)

Names of units and prefixes

1. Unit names begin in lowercase, except at the beginning of a sentence or in titles or headings in which all main words are capitalized; that is, conventional grammatical rules apply to *names*. Units named after individuals[a] also begin in lowercase. (The "degree Celsius" might appear to be an exception, but "degree" begins in lowercase and is modified by "Celsius," the name of an individual. Use of "degrees centigrade" is obsolete.)
2. Apply only one prefix to a unit name (e.g., nm, not mμm). The prefix and unit name are joined without a hyphen or space between. In three cases, the final vowel of the prefix is dropped: megohm, kilohm, and hectare. Prefixes are added to "gram," not to the base unit "kilogram." Prefixed are never used by themselves.
3. If a compound unit involving division is spelled out, the word *per* is used (not a slash or solidus, except in tables in which space may be limited). Only one *per* is permitted in a written unit name (see Rule 30 below).
4. If a compound unit involving multiplication is spelled out, the use of a hyphen is usually unnecessary, but it can be used for clarity (e.g., newton meter or newton-meter). The multiplication (product) dot (·) should not be used when unit names are spelled out.
5. Plurals of unit names are formed by adding an "s," except that hertz, lux, and siemens remain unchanged, and henry becomes henries.
6. Names of units are plural for numerical values greater than 1, equal to 0, or less than -1. All other values take the singular form of the unit name. Examples: 100 meters, 1.1 meters, 0 degrees Celsius, -4 degrees Celsius, 0.5 meter, -0.2 degree Celsius, -1 degree Celsius, 0.5 liter.
7. NIST SP 811 (Taylor, 1995) recommends that written names of units be avoided most of the time; unit symbols should be used instead. It is appropriate, however, to use a written name the first time the unit appears in a text if it is felt that readers might not be familiar with the unit.

Symbols for units

8. Unit symbols should be thought of as mathematical entities: The physical quantity equals the numeral multiplied by the value represented by the unit symbol. Hence, with few exceptions (see Rules 7 and 15), symbols are used when units are used in conjunction with numerals.
9. Written symbols are never made plural (that is, by addition of "s").
10. A symbol is not followed by a period except at the end of a sentence.
11. Symbols for units named after individuals[a] have the first letter capitalized, but the name of the unit is written in lowercase (see rule 1). Other symbols are not capitalized except that the second level authorities recommend a capital L instead of a lower case l as the symbol for the liter to avoid confusion with the numeral one (1). Both L and l are recognized by the primary authority as symbols for the liter. The capital L is recommended here.
12. Symbols for prefixes greater than kilo are capitalized; kilo and all others are lowercase. It is important to follow this rule because some letters for prefixes are the same as some symbols or another prefix: G for giga and g for gram; K for kelvin and k for kilo; M for mega and m for milli and for meter; N for newton and n for nano; and T for tera and t for metric ton.

Continued

Table 4. Summary of SI Style Conventions (Rules) (continued)

13. Use numerical superscripts (^{2} and ^{3}) to indicate squares and cubes; do not use sq., cu., or c. It is also better, when unit names are written out, to use the form "second squared" rather than "square second" unless volume or area are being discussed: "square meter," "cubic meter."
14. Exponents also apply to the prefix attached to a unit name; the multiple or submultiple unit is treated as a single entity. Thus μm^3 is the same as 10^{-18} m^3.
15. Third-level sources and English style manuals recommend that sentences should not begin with numerals. Because a unit symbol is always preceeded with a number (numeral), a sentence can never begin with a unit name or symbol. Whenever possible, a writer should recast a sentence so it does not begin with a numeral; if that can't be done, the number and unit name should be spelled out.
16. Compound symbols formed by multiplication may contain a product dot (·) to indicate multiplication; international rules say that this may be replaced with a period or a space. In the United States, the product dot is recommended. Compound symbols formed by division can use a slash (/), a horizontal line with units above and below, or be indicated by negative exponents; e.g., $\mu mol \cdot m^{-2} \cdot s^{-1}$, $\mu mol \cdot mol^{-1}$, etc. In no case should symbols be run together (e.g., Wm^{-2}).
17. Because compound unit symbols are mathematical entities, they must not include nonsymbol words or abbreviations. This is not true of unit names without numerals. Thus an author must avoid "μmol CO_2 (mol of air)$^{-1}$" but can write: "Data are presented as micromoles of CO_2 per mole of air ($\mu mol \cdot mol^{-1}$)." (See discussion in the text.)
18. Do not mix symbols and spelled-out unit names (e.g., W per square meter), and *never* mix SI units or their accepted relatives (e.g., liter, minute, hour, day, plane angle in degrees) with units of another system such as the CGS or the English system (e.g., miles per liter, kg ft^{-3}, or grams per ounce for the quantity of fat in a food).
19. The percent symbol (%) is an acceptable unit for use with the SI: % = 0.01. When used, a space is left between the symbol % and the number by which it is multiplied: $X = 25\ \% = 25 \times 0.01 = 0.25$. Rather than using such terms as "percentage by volume" (meaningless because % is simply a number), a recommended approach is to present data as mL/L, µmol/mol, g/kg, mol/L, mol/kg, etc. (Taylor, 1995).
20. Unit symbols are printed in roman type (upright letters); *italic letters* (slanted) are reserved for quantity symbols, such as A for area, m for mass, t for time, and Ψ for water potential. For typewriting or longhand, underlining may be used as a substitute for italics. According to this rule, the Greek mu, µ, when used as the prefix symbol for micro, should be printed in roman type (not in italics).

Numerals, often with Symbols

21. A space is left between the last digit of a numeral and its unit symbol. A product dot (·), space, or slash (/) is used between unit symbols when more than one is used; see rule 16. Exceptions are the degree, minute, and second symbols for angles or latitudes (e.g., 30° north). Note that the degree Celsius (°C) is a single unit symbol (no space between ° and C) that should also be preceeded by a space. It is incorrect to use 12 ° to 25 °C (that is, to use ° without C); correct forms are: 12 °C to 25 °C, (12 to 25) °C, or 12–25 °C.
22. When a quantity is used in an adjectival sense, English rules of grammar suggest that a hyphen should be used between the numeral and the unit *name:* a five-hundred-watt lamp. But when unit *symbols* are used, the hyphen should be omitted: a 500 W lamp (because the symbol is a mathematical entity, and the hyphen could be mistaken for a minus sign).
23. In the United States, the period is used as the decimal marker although some countries (e.g., France, Germany, Great Britain) use a comma or a raised period.

Continued

Table 4. Summary of SI Style Conventions (Rules) (continued)

24. To avoid confusion (because some countries use a comma as a decimal marker), a space should be used instead of a comma to group numerals into three-digit groups; this rule may be followed to the right as well as to the left of the decimal marker. Omission of the space is preferred when there are only four digits, unless the numeral is in a column with others that have more than four digits. (In spite of this rule, many journals that consistently use a period as the decimal marker also use the comma to group numerals into three-digit groups.)
25. Decimal fractions are preferred to common fractions.
26. Decimal values less than one have a zero to the left of the decimal (e.g., 0.2 m).
27. Multiples and submultiples are generally selected so that the numeral coefficient has a value between 0.1 and 1000. Exceptions occur when the differences between numbers being compared are extreme (e.g., 1500 m of 2 mm wire), and for comparison, especially in tables, similar quantities should use the same unit, even if the values fall outside this range.
28. With numerals, do not substitute the product dot (·) for a multiplication sign (×). (E.g., use 2 × 2, not 2·2.)

The denominator

29. For a compound unit that is a quotient, use "per" to form the name (e.g., meters per second) and a slash (/; solidus) to form the symbol, with no space before or after the slash (e.g., m/s). Compound units may be written with negative exponents (e.g., $m{\cdot}s^{-1}$ or $m\ s^{-1}$).
30. Do not use two or more "pers" or slashes in the same expression because they are ambiguous (see Rule 3); negative exponents avoid this problem: $J{\cdot}K^{-1}{\cdot}mol^{-1}$ (not J/K/mol); J/K·mol is acceptable because all symbols to the right of the slash belong to the denominator.
31. Many third-level sources suggest that the denominator should not be a multiple or submultiple of an SI base unit (e.g., $\mu N{\cdot}m^{-2}$ but not $N{\cdot}\mu m^{-2}$). (But see discussion in the text.)

[a] Individuals after whom units are named include: Antoine Henri Becquerel (France, 1852-1908), Anders Celsius (Sweden, 1701-1744), Charles Augustin de Coulomb (France, 1736-1806), Michael Faraday (England, 1791-1867), Heinrich Rudolf Hertz (Germany, 1857-1894), James Prescott Joule (England, 1818-1889), Lord William Thomson Kelvin (Scotland, 1824-1907), Sir Isaac Newton (England, 1643-1727), Georg Simon Ohm (Germany, 1787-1854), Blaise Pascal (France, 1623-1662), Sir William Siemens (Germany, Great Britain, 1823-1883), Count Allessandro Giuseppe Antonio Anastasio Volta (Italy, 1745-1827), and James Watt (Scotland, England, 1736-1819).

The CIPM recognized in 1969 that users of SI will also wish to employ with it certain units that are not part of it, but that are important and are widely used. These units (along with the unified atomic mass unit and the standard acceleration due to gravity), are shown in Table 5. Note that a goal in setting up the International System of Units was to produce a coherent system, as noted above, a system in which derived units are various combinations of the base units without the necessity of including numerical multiplication factors. All of the units in Table 5 do require the use of such factors, and hence they lose the advantages of the coherence of SI units. It was recommended that their use be restricted to special cases. It is clear, however, that plant scientists will use the minute, hour, and day (not to mention the week, month, and year) without hesitation in reporting methods and results. The liter is also a much more convient unit for plant scientists than the cubic meter, which is the official SI unit of volume. Thus, we can be thankful for the CIPM's decisions in 1969 and for Table 5!

Table 5. Some Units used with the SI but not Officially Part of SI[a]

Name	Symbol	Value in SI units
minute	min	1 min = 60 s
hour	h	1 h = 60 min = 3600 s
day	d	1 d = 24 h = 86 400 s
degree	°	$1° = (\pi/180)$ rad
minute	′	$1' = (1/60)° = (\pi/10\,800)$ rad
second	″	$1'' = (1/60)' = (\pi/648\,000)$ rad
liter (litre)	L (l)	$1\ \text{L} = 1\ \text{dm}^3 = 10^{-3}\ \text{m}^3$
metric ton (tonne)	t	$1\ \text{t} = 10^3$ kg
unified atomic mass unit[b]	u	1 u = (1/12) of the mass of an atom of the nuclide ^{12}C
standard acceleration due to gravity[c]	g_n	$9.806\,65\ \text{m}\cdot\text{s}^{-2}$

[a] Because these units must be multiplied by a factor to make them equivalent to SI units, they are not coherent in the sense of other SI units.

[b] The actual value of the unified atomic mass unit in SI units must be determined by experimentation. At present it is considered to be: $u = 1.660\,540\,2(10) \times 10^{-27}$ kg. The uncertainty of the last two figures, at the level of one standard deviation, is shown in parentheses.

[c] This value was confirmed in 1913 by the 5th CGPM. Its symbol, g_n, should be used instead of the many symbols currently used to indicate one acceleration due to gravity at the earth's surface (e.g., g, *g*, G, G, ×g, etc.).

Table 6 includes some units that were used with the metric system but that the CIPM recommends should not be used with the SI. A few of these units continue to be in wide use among plant scientists.

3. SOME SPECIAL CONSIDERATIONS

Although most of the rules in Table 4 are explained adequately in the table, a few of them as well as some of the units in Tables 5 and 6 are worthy of discussion.

A. Language Conventions with SI-Unit Names and Symbols.[2] Written out names for the units follow the rules of grammar (English or other language), whereas the unit *symbols* should be thought of as mathematical entities by which the preceeding numeral is multiplied. For example, unit *names* begin with a lowercase letter unless grammatical rules call for uppercase (i.e., at the beginning of a sentence and in titles), but the upper or lowercase of *symbols* must never be changed regardless of where they appear. In English, names are often made plural

[2] Rules in Table 4 that bear on this discussion are: 1, 7, 8, 15, 17, & 22.

by addition of *s*, but SI symbols never are. Numbers (usually written out) followed by unit names used in an adjectival sense can be connected with a hyphen (e.g., a fifty-watt lamp) but the hyphen is not used with symbols (a 50 W lamp). Because of the mathematical nature of symbols, it is desirable to use them instead of names. Of course the name can be used the first time it appears if the reader might not be familiar with the unit or its symbol.

In some languages it is not uncommon for a numeral to begin a sentence; in English this should be avoided, preferably by recasting the sentence, but if necessary by writing out the numeral and its unit name.

Table 6. Some Discarded Metric Units

Discarded Metric Unit	Acceptable SI Unit
micron (μ)	micrometer (μm)
millimicron (mμ)	nanometer (nm)
ångstrom[a] (Å)	0.1 nanometer (nm)
bar[a] (bar)	0.1 megapascal (MPa); 100 kilopascal (kPa)
calorie (cal)	4.1842 joule (J)
degree centigrade (°C)	degree Celsius (°C)
hectare[a] (ha)	10 000 m^2 or 0.01 km^2
einstein (E)	mole of photons or quanta (mol)
dalton (Da)	unified atomic mass unit (u) (see Table 5)
standard "gravity" (g, *g*, G, **G**, ×g, etc.)	standard acceleration due to gravity (g_n)
molar solution (M)	$mol \cdot L^{-1}$ ($kmol\ m^{-3}$)
molal solution (m)	$mol \cdot kg^{-1}$
parts per million (ppm)	$mg \cdot kg^{-1}$ $\mu mol \cdot mol^{-1}$ (e.g., CO_2 in air) (Use kg for mixed substances and mol for pure substances and gases.) 1000 $mm^3 \cdot m^{-3}$ (volume; e.g., liquids)
parts per billion (ppb)	$\mu g \cdot kg^{-1}$ $nmol \cdot mol^{-1}$ $mm^3 \cdot m^{-3}$ (volume; e.g., liquids)

[a] In view of existing practice in certain fields, the CIPM (1978) considered that these units could be used with the SI temporarily although they should not be introduced where they are not used at present.

Because names follow grammatical rules, it is acceptable to use them in conjunction with other terms, but such terms must *not* be included with SI units: "Photon flux was measured as moles of photons in the photosynthetically active range (400 to 700 nm) per square meter second ($\mu mol \cdot m^{-2} \cdot s^{-1}$)." "Data are presented as milligrams of protein per gram of fresh tissue (mg/g)." *Or:* "Protein data are presented on a fresh-mass basis ($mg \cdot kg^{-1}$)." This rule is often overlooked by plant physiologists, who even sometimes construct meaningless symbols to present their data: mg/gfw (meaning milligrams per gram of fresh weight) *or* mg (kg fresh mass)$^{-1}$. The rule was discussed by Downs (1988) and perhaps in other third level sources of which I am not aware, but it has otherwise been largely overlooked by plant physiologists. *The rule is emphasized, however, by the second-level authorities* (ISO Standards Handbook, 1993; NIST SP 811, Taylor, 1995). Plant scientists should improve the rigor of their presentations by adhering to this rule.

B. Space Between Numerals and Units and Within Compound Units. For some unknown reason it has become increasingly common to omit the space between a numeral and the unit that follows (e.g., a 50mL flask). In the worst cases, the space or product dot is omitted between symbols in a multiple unit, creating new symbols that have no meaning (e.g., Wm^{-2}, $\mu molm^{-2}s^{-1}$, etc.). This practice breaks Rules 16, 19, and 21 in Table 4, and leaving out the space can confuse readers. Use of the product dot is highly recommended in the United States (Taylor, 1995), but plant scientists have used it only infrequently. Its consistent use would remove any ambiguity from multiple units and would overcome the tendency to run units together.

C. Italics for Quantity Symbols, Roman for Unit Symbols. This simple practice is stated in Rule 20, Table 4, but many plant scientists seem to be unaware of it. Remember that unit symbols are printed in roman type (upright letters); *italic letters* (slanted) are reserved for quantity symbols, such as A for area, m for mass, t for time, and Ψ for water potential. For typewriting or longhand, underlining may be used as a substitute for italics. According to this rule, the Greek mu, μ, when used as the prefix symbol for micro, should be printed in roman (i.e., upright) type whenever possible. Unfortunately, not all word processors allow this. (Note that Greek, Roman, or even Cyrillic *alphabets* can be printed in either roman or *italic* type).

D. Only One per or Slash in a Multiple Unit. This is another simple rule that plant scientists should apply more widely (Rule 30, Table 4): Do not use two or more "pers" or slashes in the same expression because they are ambiguous. Negative exponents avoid this problem: $J \cdot K^{-1} \cdot mol^{-1}$ (not J/K/mol); J/K·mol is acceptable because all symbols to the right of the slash belong to the denominator. If this is written out, it becomes: joules per kelvin mole.

E. Only Base Units in Denominators. As noted in Rule 31, Table 4, many third-level sources suggest that the denominator should not be a multiple or submultiple of an SI base unit (e.g., $\mu N \cdot m^{-2}$ but not $N \cdot \mu m^{-2}$). As editor of Journal of Plant Physiology during the past six years, I have found this rule to be the most difficult to enforce. It goes against much tradition and sometimes seems illogical and unreasonable. For example, authors have long reported amounts or concentrations of metabolites, hormones, and other compounds as μg/mg, nmol/mL, etc. It may seem to go against one's intuition to use the equivalents of those two examples: g/kg, μmol/L (or, using "true" SI units: mmol/m^{-3}).

Now I learn that the rule of only base units in denominators is recommended solely in third-level sources but is not an official SI rule and is not in the ISO Standards Handbook or NIST SP 811 (Taylor, 1995), which are second-level sources with virtually as much authority as the primary SI publication. It is not necessary to adhere to every suggestion put forth in third-level publications. Thus, I will no longer try to enforce the rule in my editing.

Having said that, I will nevertheless make an argument (third level!) that the rule (suggestion, at least) can, in many cases, be quite logical and helpful. Use of a single denominator for a given quantity by everyone in the field allows us to think about the various quantities without the necessity of mentally converting them to *our* familiar range, different perhaps from that of the author we are reading. Of course, to take advantage of this rule, those within the field must agree to form *new* habits. For example, photosynthesis rates were previously expressed on the basis of CO_2 uptake (as milligrams or moles) per square decimeter of leaf surface, perhaps because a square decimeter seemed to be an area similar to the area of real leaves. Most workers now express most parameters relating to photosynthesis on the basis of a square meter, as suggested by this rule. After all, few real leaves are exactly 1.0 dm^2 or exactly 1.0 m^2 in area. (Does a banana leaf approach 1.0 m^2?) And irradiance is commonly expressed on the basis of a square meter: $\mu mol \cdot m^{-2} \cdot s^{-1}$ or $W \cdot m^{-2}$. When everything is expressed on the same basis (m^{-2}), comparisons are much easier.

In spite of the traditions noted above, units for quantities of mass can easily follow the rule. Why express the amount of growth regulator in a tissue sample as 8.5 pg mg^{-1} when it is just as easy to write 8.5 µg kg^{-1}? A kilogram is a relatively large amount of tissue, but it is easy enough to visualize. The researcher probably didn't use a 1.0 mg sample of tissue any more than he or she used a 1.0 kg sample. Uniform adherence to this rule would soon familiarize researchers with its merits. Nevertheless, applying the rule is optional.

Because a cubic meter is so large, it may seem a little less logical to express solution concentrations on the basis of a cubic meter, which is the SI base unit of volume. Nevertheless, many plant scientists have decided to use the cubic meter as the base unit for solution concentrations: 1.0 m^3 = 1000 L; thus a 1.0 mmol/L solution = 1.0 mol m^{-3}. Because SI rules allow use of the liter, however, even though it is not an official part of the system, and because concentrations based on the liter have long been used by plant physiologists (and many solutions *are* made up in liter quantities), it is acceptable to use liters in most journals that publish papers in the plant sciences. (See discussion below on molarity and molality; their traditional units, M and m, should be phased out of use by plant scientists.)

It is not always possible or desirable to have only base units in denominators. For example, spectral energies *must* specify a narrow wave-length range, the nanometer: $mol \cdot m^{-2} \cdot s^{-1} \cdot nm^{-1}$ or $W \cdot m^{-2} \cdot nm^{-1}$. (The range that was actually measured should always be stated in the methods section)

In some cases it may be preferable to write out information for further clarity. For example, a strict editor trying to enforce this rule would insist that a temperature gradient of 1 K mm^{-1} be written as 1000 K m^{-1}. It would be better to state: "A temperature gradient of 1 K over a distance of 1 mm was measured."

The recommendation of this third-level publication: *When it is logical and helpful to do so, use only SI base units in denominators.*

F. The Liter: Symbols and Use; Molar and Molal Solutions. The liter (litre in England, France) is not an official part of SI, probably because it is not "coherent."

To derive it from the cubic meter, the SI base unit, a multiplication factor must be used (1 L = 0.001 m^3). The 12th CGPM in 1964 declared, however, "that the word 'liter' may be employed as a special name for the cubic decimeter" and recommended "that the name liter should not be employed to give the results of high accuaracy volume measurements." (Taylor, 1993.) These statements effectively defined the liter as exactly 1 dm^3 and the milliliter (mL) as 1 cm^3. Because the liter has a convient size and the term is traditional and widely used by non-scientists, plant scientists will continue to use it and the milliliter. Even the deciliter might sometimes be most convenient. Of course, we could use dm^3 and cm^3 as easily.

In 1979, the CGPM considered "that, in order to avoid the risk of confusion between the letter l and the number 1, several countries have adopted the symbol L instead of l for the unit liter..." It was further decided "to adopt the two symbols l and L as symbols to be used for the unit liter, considering further that in the future only one of these two symbols should be retained..." NIST SP 811 (Taylor, 1995) strongly recommends L as the symbol for the liter.

The vast majority of papers in plant physiology express concentrations in terms of **molarity** (symbol M = $mol \cdot L^{-1}$) or, especially in the field of water relations, of **molality** (symbol m = $mol \cdot kg^{-1}$). Nevertheless, both second-level authorities (ISO Standards Handbook and NIST SP 811, Taylor, 1995) recommend that the symbols for these terms be discontinued. (Although not stated in those sources, the terms themselves might still be used.) The reason is that those unit symbols (M and m) are specialized symbols that might not be understood by someone outside the field (e.g., a physicist), whereas $mol \cdot L^{-1}$ and $mol \cdot kg^{-1}$ are simple SI units understood by anyone familiar with SI. Furthermore, m for molality might be mistaken as m for meter. This recommendation is confirmed by an important third-level source, *Quantities, Units and Symbols in Physical Chemistry* (Mills et al., 1995). Most plant physiologists will no doubt continue to use the terms *molarity* and probably *molality*, but it is recommended that the equivalent SI units be used instead of the traditional symbols. (But see Table 1 in Chapter 10.)

G. The Dalton and the Unified Atomic Mass Unit. Many biochemists and most (virtually all) plant physiologists use the **dalton** (Da or D) as a unit of atomic or molecular mass. The dalton has, however, never been accepted by the CGPM, and it is exactly equivalent to the **unified atomic mass unit** (symbol u, Table 5), which has been considered and accepted by the CGPM and is published in the first-level authority. Hence, there seems to be little excuse beyond tradition to use the dalton. Most plant scientists gave up the einstein in favor of mole of photons. The recommendation is to begin to use the unified atomic mass unit with its symbol u, probably with some explanation until it becomes more familiar to plant scientists.

H. Equivalent of Gravity at the Earth's Surface. It is common for biochemists and others to express the acceleration experienced by a sample being centrifuged as multiples of the average acceleration caused by gravity at the earth's surface. There has been almost no agreement, however, on the symbol that should be used for this value. On sees in various publications: G, g, *g*, **G**, **g**, ×g, and no doubt others. The problem with these symbols is that g is the symbol for gram, G is the prefix symbol for giga, and italics (*g*) is reserved for physical quantities rather than units. **Bold facing** has no precident in the use of units. Actually, there never should have been a problem because the CGPM established the *standard acceleration due to gravity* in 1901 and confirmed the value in 1913. The primary and secondary sources show the symbol g_{n}. The logic of this symbol is that the *acceleration of free fall* (g) is a physical quantity (hence *italics*) that can have any value (units: $\mathrm{m \cdot s^{-2}}$), but the *standard acceleration of free fall* (indicated by the subscript, which is not in italics: $g_{\mathrm{n}} = 9.806\ 65\ \mathrm{m\ s^{-2}}$) is the value of particular interest. It must be experimentally determined and is thus a noncoherent unit, but multiples of this value can be used to describe the acceleration caused by centrifugation (e.g., sample centrifuged for 20 min at 1000 g_{n}) or the acceleration experienced in an orbiting satellite (e.g., $10^{-3}\ g_{\mathrm{n}}$). This symbol, in context, should be readily understood by everyone without any special explanation.

I. Other Discarded Metric Units. Table 6 includes a number of discarded metric units that have not yet been discussed. One occasionally sees *micron*, but most of us now use the *micrometer* and *nanometer*. The *ångstrom* is seldom used in the plant sciences but is still used in certain fields, which is permitted according to the footnote in Table 6. The *bar* is still used in meteorology and sometimes in the field of plant water relations, but its replacement with the coherent *megapascal* or *kilopascal* has been accepted by most plant physiologists. The *hectare* (ha) will no doubt continue to be used by agriculturists instead of the more correct $\mathrm{hm^2}$ or $\mathrm{m^2}$. The *Calorie* (kilocalorie) is a part of our modern dieting culture, but (in the United States) so is the Fahrenheit temperature scale; scientists use degrees Celsius and should also use *joules* instead of calories. One often sees *parts per million* (or *billion* or even *trillion*), but it is more logical to use their equivalents in units of mass, volume, or amount of substance (e.g., $\mathrm{mg \cdot kg^{-1}}$, $\mathrm{mmol \cdot kg^{-1}}$, $\mathrm{mol \cdot L^{-1}}$). With such units, it is not necessary to specify the basis of comparison (i.e., volume, mass, etc.).

IMPORTANT REFERENCES FOR APPLICATION OF SI UNITS

Many of these publications are now out of date and are included here only for historical reference. The most recent and most recommended publications that have come to my attention are written in **bold face**.

American National Metric Council. 1993. ANMC Metric Editorial Guide, Fifth Edition. American National Metric Council, 4330 East/West Highway, Suite 1117, Bethesda, MD 20814.

[Anonymous]. Standard Practice for Use of the International System of Units. ASTM E380-89. American Society for Testing and Materials, 1916 Race Street, Philadelphia PA 19103. [No date.]

[Anonymous]. 1992. *Guidelines for measuring and reporting environmental parameters for plant experiments in growth chambers.* ASAE Engineering Practice: ASAE EP411.1. American Society of Agricultural Engineers, 2950 Niles Road, St. Joseph, Michigan 49085-9659. [This is Appendix C in this book.]

[Anonymous]. 1979. Metric Units of Measure and Style Guide. U. S. Metric Association, 10245 Andasol Avenue, Northridge CA 91103.

[Anonymous]. 1985. *Radiation quantities and units.* ASAE Engineering Practice: ASAE EP402. American Society of Agricultural Engineers, 2950 Niles Road, St. Joseph, Michigan 49085-9659.

[Anonymous]. 1982. SI Units Required in Society Manuscripts. Agronomy News (March-April 1982, p 10-13).

[Anonymous]. 1988. *Use of SI (metric) units.* ASAE Engineering Practice: ASAE EP285.7. American Society of Agricultural Engineers, 2950 Niles Road, St. Joseph, Michigan 49085-9659.

Boching, P.M. 1983. Author's Guide to Publication in Plant Physiology Journals. Desert Research Institute Pub. No. 5020. Reno, Nev.

Buxton, D.R., and D.A. Fuccillo. 1985. Letter to the editor. Agronomy Journal 77:512-514. [This letter includes a summary of a survey of 97 journals; 77 percent either required or encouraged the use of SI units.]

Campbell, G.S., and Jan van Schilfgaarde. 1981. Use of SI units in soil physics. Journal of Agronomic Education. 10:73-74.

CBE Style Manual Committee. 1994. Scientific style and format: the CBE manual for authors, editors, and publishers. 6th edition. Cambridge University Press, Cambridge, New York. [See also earlier editions of CBE Style Manual.]

Downs, Robert J. 1988. Rules for using the International System of Units. *HortScience* 23: 811-812.

Goldman, David T., and R.J. Bell, editors. 1986. The International System of Units (SI). National Bureau of Standards Special Publication 330. U. S. Department of Commerce/National Bureau of Standards. [See Taylor (1991) for the most recent version of this publication.]

Incoll, L.D., S.P. Long, and M.R. Ashmore. 1977. SI units in publications in plant science. Current Advances in Plant Sciences 9(4):331-343. [This article recommended several practices that are now in wide use by plant scientists. It was a kind of historical turning point.]

ISO Standards Handbook. 1993. *Quantities and Units.* International Organization for Standardization, Genève. [This is the highly authorative, second-level reference. It is available from American National Standards Institute, 11 West 42nd Street, New York, NY 10036.]

Mills, Ian, Tomislav Cvitaš, Klaus Homann, Nikola Kallay, and Kozo Kuchitsu. 1995. *Quantities, Units and Symbols in Physical Chemistry 2nd Edition.* Blackwell Scientific Publications, Oxford, London, Endinburgh, Boston, Palo Alto, & Melbourne.

Monteith, J.L. 1984. Consistency and convenience in the choice of units for agricultural science. Experimental Agriculture. 20:105-117.

Petersen, M.S. December 1990. Recommendations for use of SI units in hydraulics. Journal of the Hydraulics Division, Proceedings of the American Society of Civil Engineers 106:HY12.

Savage, M.J. 1979. Use of the international system of units in the plant sciences. HortScience 14:493-495.

Salisbury, F.B. 1991. *Systèm Internationale:* The use of SI units in plant physiology. Journal of Plant Physiology 139(1):1-7.

Taylor, Barry N., editor. 1991. *The International System of Units (SI).* National Institute of Standards and Technology Special Publication 330. U.S. Government Printing Office, Washington, D.C. [This is the United States edition of the English translation of the sixth edition of "Le Systèm International d'Unités (SI)", the definitive publication of the International Bureau of Weights and Measures and thus the first-level authority. There is also a British version with slight differences, as in the spelling of "metre," "litre," and "deca." The United States version is for sale by the Superintendent of Documents, U. S. Government Printing Office, Washington, DC 20402.]

Taylor, Barry N. 1995. *Guide for the Use of the International System of Units (SI).* National Institute of Standards and Technology Special Publication 811. [Along with the ISO Standards Handbook, this publication should be considered second in authority only to *"Le Systèm International d'Unités (SI),"* at least for citizens of the United States.]

Thien, S.J., and J.D. Oster. 1981. The international system of units and its particular application in soil chemistry. Journal of Agronomic Educucation 10:62-70.

U.S. Metric Association. 1993. Guide to the Use of the Metric System [SI Version]. U.S. Metric Assocation, Inc., 10245 Andasol Avenue, Northridge, CA 91325-1504.

Vorst, J.J., L.W. Schweitzer, and V.L. Lechtenberg. 1981. International system of units (SI): Application to crop science. Journal of Agronomic Educucation 10:70-72.

Weast, Robert C., editor. (1995 and new editions each year). CRC Handbook of Chemistry and Physics. CRC Press, Boca Raton, Fla.

CONSULTANTS

Bruce G. Bugbee
Utah State University
Logan, Utah

Louis Sokol*
Boulder, Colorado

John Sager
Kennedy Space Center, Florida

Barry N. Taylor*
Nat. Inst. of Standards & Technology
Gaithersburg, Maryland

*Dr. Sokol is president emeritus of the U.S. Metric Association and a member of the National Conference on Weights and Measures. He is a certified metrication specialist. Dr. Taylor is the U. S. representative on the CGPM.

2

RULES FOR BOTANICAL NOMENCLATURE

John McNeill
Royal Ontario Museum
100 Queen's Park
Toronto, M5S 2C6, Canada

and

Mary E. Barkworth
Biology Department
Utah State University
Logan, Utah 84322-5300, U.S.A.

The following discussion provides some recommendations for documenting the plant material used in experimental and other studies and summarizes the rules of nomenclature that have been established at botanical congresses held every five or six years for over a century (for the most recent edition of the rules, see Greuter et al., 1994).

1. DOCUMENTATION

It is imperative that the plant or fungal material used in any experiment be documented. The source of the seeds, plants, or cultures used should be cited in the publication, either by indicating the supplier (e.g., commercial source, culture collection) and including any cultivar or strain identification, or else, in the case of material obtained from the wild, by a statement of the precise geographical location. In addition, in comparative studies, or in those in which the material would be difficult or impossible to replicate (e.g., plants obtained from most wild sources), representative material should be deposited in a recognized herbarium or culture collection, as appropriate. The herbarium specimens should include plants at reproductive maturity plus representative material of any other stages used in the study. If growing the plants to the reproductive stage is not feasible, then material from as mature a plant as possible should be used. The name and location of the herbarium or culture collection where the specimens have been deposited should be reported. This can be done concisely by using the internationally accepted abbreviations given in *Index herbariorum* (Holmgren et al., 1990), or in the *World Directory* for culture collections (Staines et al., 1986). The Curator of your institutional herbarium will be able to provide advice on how to collect and preserve plant material for deposition in the herbarium. Useful advice can also be found in Fosberg & Sachet (1965), Lee et al.(1982), Radford (1986), Savile (1973), and Smith (1971), or for fungi, in Hawksworth (1974).

2. TAXONOMIC GROUPS (TAXA; SINGULAR: TAXON)

All plants are assigned to **species**, the species to **genera** (sing. **genus**), and the genera to **families**. Although differences of opinion as to circumscription of some species, genera, and families exist among taxonomists, there is good agreement on their limits for most flowering plants. Within some species, infraspecific taxa may be recognized, usually at the ranks of **subspecies** (subsp.) or **variety** (var.). The major ranks above the family (Latin: *familia*) in ascending order are: **Order** (*ordo*), **Class** (*classis*), **Division** (*divisio*) or **Phylum**, and **Kingdom** (*regnum*). Classifications at these levels are more controversial, and reference to them is not usually necessary in physiological research publications, unless the research involves comparison of a broad spectrum of plants. In further discussion of nomenclatural practice, we shall, therefore, consider only taxa at the rank of family and below. A fuller account for the general biologist of the use of scientific names of plants is to be found in Gledhill (1985).

3. FORM OF SCIENTIFIC NAMES

A. Family names. Family names are plural nouns. They should be written out in full, with the initial letter capitalized, but they are usually not italicized or otherwise set out from the rest of the text in publications from English-speaking countries. Family names, apart from nine exceptions, are based on the stem of a generic name to which the suffix -aceae is attached. Eight of the exceptions are simple alternatives: Cruciferae [alternative based on genus = Brassicaceae], Compositae [= Asteraceae], Gramineae [= Poaceae], Guttiferae [= Clusiaceae], Labiatae [= Lamiaceae], Leguminosae [= Fabaceae] (but see below), Palmae [= Arecaceae], and Umbelliferae [= Apiaceae]. The alternative names may be used instead of the standard form but need not be. It is, however, desirable to be consistent within a paper (i.e., do not use, for example, Poaceae and Leguminosae in the same paper).

The ninth alternative name, Papilionaceae, can be used for the papilionoid legumes if they are regarded as a family distinct from the caesalpinioid and mimosoid legumes. The standard forms for these three groups of legumes, if each is regarded as a family, are Fabaceae [= Papilionaceae], Caesalpiniaceae, and Mimosaceae. The name Leguminosae (Fabaceae in the broad sense) cannot be used if these three units are treated as distinct families.

B. Names of genera. **Generic names** are composed of a single word. They should be *italicized*, underlined, or set off in some other way from regular text (e.g., written in roman if the text is italicized), and have the initial letter capitalized. They are singular nouns, not adjectives. They should be written out in full unless they are used in combination with a specific epithet as the name of a species (see next item).

C. Names of species. The name of a species is a binomial. It consists of the name of the genus followed by a single **specific epithet.** The epithet (called the "species name" in zoology) may be hyphenated but is never two separate words. It may be an adjective, or a noun in apposition, or in the genitive. The entire binomial should be italicized or set off in some other way from the main text. Underlining is commonly used when italics are not available.

As noted above, the initial letter of the generic name must be capitalized. The initial letter of the specific epithet should not be capitalized, although capitalization is permitted if: (a) the epithet is derived from the name of a person (e.g., *Plantago Tweedyi*), (b) it is derived from a vernacular name (e.g., *Dolichos Lablab*), or (c) it was once a generic name (e.g., *Picea Abies*). Use of lowercase is, however, recommended in all cases; it is never incorrect.

When writing the name of a species, both the generic name and the specific epithet must be given. The generic name may be abbreviated to the initial letter followed by a period unless it is being used for the first time in a text or there is a possibility of confusion because two genera under discussion have the same initial letter. In the latter case, a unique abbreviation consisting of the initial letter and one or more others is sometimes used for each of the generic names concerned.

D. Names of infraspecific taxa. The name of an infraspecific taxon is a combination of four words, the generic name, the specific epithet, the term denoting infraspecific rank, and the (final) infraspecific epithet. (It is possible, although uncommon, to have a hierarchy of infraspecific ranks; e.g., a subspecies with several varieties.) The term denoting rank (e.g., var., subsp.) should be in the same font as the bulk of the text, but the other words are italicized (or set off in some other way from the main text), e.g., *Stipa nelsonii* subsp. *nelsonii*, *Phyllerpa prolifera* var. *firma*, *Trifolium stellatum* f. *nanum*. As with specific epithets, lowercase should be used for infraspecific epithets (but see "Names of Cultivars" below).

E. Citation of Authorities. To be accurate and complete, it is necessary to indicate the names of the author(s) who first validly published a given name or combination. This is usually done by citing them the first time the plant or fungus name is used in the text, after which the name may be used without citation of authors. Alternatively, in papers treating many species from a particular area, a statement to the effect that the scientific nomenclature follows that used in a well known Flora or Manual for the area is generally acceptable and may be more informative. In addition it is not necessary to cite the authors of genera or taxa of higher rank unless a paper specifically addresses the taxonomy of higher ranks. Likewise, for infraspecific ranks, it is not generally necessary to cite the authorship of the species name, e.g., *Stipa nelsonii* subsp. *dorei* Barkworth & Maze. The only exception to this is when the final infraspecific epithet is the same as that of the specific epithet (so-called autonyms). In such instances, the author of the species name is given, e.g., *Stipa nelsonii* Scribner subsp. *nelsonii*. As scientific names are in Latin (or treated as Latin), the ampersand (&) or the Latin word 'et' should be used when more than one author is involved, never the word 'and' or its equivalents in other modern languages. Parentheses indicate that the taxon was originally named in another genus or at a different rank but with the same (final) epithet.

The name of the person who first validly published the combination being used appears to the right of the parenthetical authors. For example, *Agropyron cristatum* (L.) Gaertner shows that the taxon was first named by Linnaeus ("L."), who coined the epithet *cristatum* for it, but that he adopted a different taxonomic treatment; in this case he included it in another genus (*Triticum*). Gaertner was the first person to combine the epithet *cristatum* with the generic name *Agropyron*. In botanical nomenclature, unlike the practice in zoology, the person publishing the combination being used, as well as the original author(s), must be cited.

If the names of authors are connected by 'ex,' it means that the author(s) named after the 'ex' were responsible for valid publication of the name, but they attributed the name to the author(s) whose name(s) precede the 'ex.' For example, *Carex stipata* Muhlenb. ex Willd. means that Willdenow published the combination but attributed it to Muhlenberg who, although he had used the epithet, had not, in fact, previously validly published it. If one wishes to abbreviate the citation, one should retain the names of those actually publishing the combination; i.e, those after the 'ex' (in this case 'Willd.'). Authors' names are sometimes connected by the word 'in'. This implies that the first person actually named the taxon and provided the description, but that it was published in a work written by the author(s) named after the 'in;' e.g., *Viburnum ternatum* Rehder in Sargent means that Rehder described and named the species, but that it was published in a larger work written by Sargent. In such circumstances, the 'in' and the name following it are not strictly part of the author citation and are better omitted unless the place of publication is being cited (in this case retaining only Rehder).

The names of authors can be abbreviated, but to avoid confusion, one should follow the abbreviations used in some standard work (e.g. *Hortus Third* [Bailey Hortorium, 1976], the *Authors of Plant Names* [Brummitt & Powell, 1992] or a major Flora of the area). Abbreviated names are followed by a period (e.g., Willd. is an abbreviation for Willdenow in the example above).

In the cases of names of fungi, ":Pers." or ":Fries" after the name of the author indicates that such names were sanctioned for use by Persoon or Fries, respectively, and have a preferred nomenclatural status (Hawksworth, 1984).

4. SPECIAL SITUATIONS

A. Names of Hybrids. A hybrid between taxa may be referred to by placing a multiplication sign × between the names of its two parental taxa; e.g., *Agrostis* L. × *Polypogon* Desf., *Polypodium vulgare* subsp. *prionodes* Roth × subsp. *vulgare*. Some hybrids have been given a name of their own. Their hybrid status is indicated by placing a multiplication sign immediately before the name, e.g. ×*Agropogon* P. Fourn. (= *Agrostis* L. × *Polypogon* Desf.), *Mentha* ×*smithiana* R.A. Graham (= *M. aquatica* L. × *M. spicata* L.) If the mathematical symbol is not available, a lower case 'x' should be used (not italicized) and a single space inserted between it and the name to promote clarity; e.g., *Mentha* x *smithiana* R.A. Graham.

B. Controversial or Unfamiliar Names. If there is controversy over the name of a taxon, or if one is using the correct but still unfamiliar name for a taxon, a familiar alternative name (synonym) should be given within square brackets (or otherwise indicated parenthetically) immediately after the first mention of the name; e.g., *Achnatherum hymenoides* (Roemer & Schultes) Barkworth [= *Oryzopsis hymenoides* (Roemer & Schultes) Ricker or *Stipa hymenoides* Roemer & Schultes]; or *Elymus lanceolatus* (Scribner & J. G. Smith) Gould [= *Agropyron dasystachyum* (Hooker) Scribner & J. G. Smith].

C. Names of Cultivated Plants. The names of cultivated plants follow the rules of nomenclature for other plants in so far as these are applicable (e.g., *Triticum aestivum* L. for the commonly cultivated species of wheat), but names of cultivated varieties or races (termed **"cultivars"**) are subject to additional rules. The name of a cultivar follows that of the lowest botanical rank to which it can be assigned. For example, cultivars of wheat would have the cultivar name given after *Triticum aestivum,* but for hybrid tea rose cultivars, which are the result of extensive interspecific hybridization, the cultivar name would follow the generic name *Rosa*.

The cultivar name is not italicized, but its initial letter is in uppercase. It should be put between single quotation marks, e.g., *Taxus baccata* 'Variegata'; until recently it could also be preceded by cv. (for cultivar), e.g., *Taxus baccata* cv. Variegata. The group of cultivars to which it belongs may also be indicated, e.g., *Rosa* (Hybrid Tea) 'Peace'.

The names of graft-chimeras consist of the names of the components, in alphabetical order, connected by the addition (plus) sign: "+" (e.g., *Cytisus purpureus* + *Laburnum anagyroides*; *Syringa* ×*chinensis* + *S. vulgaris*). For further information on the names of cultivated plants, see Trehane et al. (1995).

D. Pleomorphic Fungi. Fungi with different phases in their life-cycle can have different names applied to their various states. The fungus in all its parts is known by the name of the sexually reproducing stage (teleomorph), but, where convenient, separate names can be used for the stages reproducing by asexual methods (anamorphs). Anamorph names make clear the phase of the fungus that has been used in physiological studies and so should be cited wherever appropriate.

E. Common Names. Common names (or specially formed names in vernacular languages; e.g., English) are permitted in most journals of plant physiology and related sciences, but the scientific name and its author(s) should always be stated in parentheses immediately following the first use of the common or vernacular name.

REFERENCES

Bailey Hortorium. 1976. Hortus third. Macmillan, New York; Collier Macmillan, London. 1290 p.

Brummitt, R.K., and C.E. Powell. 1992. Authors of Plant Names. Royal Botanic Gardens, Kew. p 732.

Fosberg, F.R., and M.-H. Sachet. 1965. Manual for Tropical Herbaria. International Bureau for Taxonomy and Nomenclature, Utrecht. p 132 (*Regnum veg.* 39).

Gledhill, D. 1985. The Names of Plants. Cambridge University Press, Cambridge & New York.

Greuter, W., F.R. Barrie, H.M. Burdet, W.G. Chaloner, V. Demoulin, D.L. Hawksworth, P.M. Jørgensen, D.H. Nicolson, P.C. Silva, P. Trehane, and J. McNeill. 1994. International Code of Botanical Nomenclature (Tokyo Code). Koeltz Scientific Books, Königstein Germany. (*Regnum veg.* 131).

Hawksworth, D.L. 1974. Mycologist's Handbook. Commonwealth Mycological Institute, Kew.
Hawksworth, D.L. 1984. Recent changes in the international rules affecting the nomenclature of fungi. Microbiological Sciences 1:18-21.
Holmgren, P.K., N.H. Holmgren, and L.C. Barnett. 1990. Index Herbariorum. Part 1. The Herbaria of the World. ed. 8. New York Botanical Garden, Bronx, New York. 693 p (*Regnum veg.* 120).
Lee, W.L., B.M. Bell, and J.F. Sutton, editors. 1982. Guidelines for Acquisition and Management of Biological Specimens. Association of Systematic Collections, Lawrence, Kansas.
Radford, A.E. 1986. Fundamentals of Plant Systematics. Harper and Row, New York. 498 p.
Savile, D.B.O. 1973. Collection and Care of Botanical Specimens. (Reprint with addendum) Publ. 1113. Agriculture Canada, Ottawa.
Smith, C.E. 1971. Preparing Herbarium Specimens of Vascular Plants. Agricultural Information Bulletin 348, Agricultural Research Service, United States Department of Agriculture.
Staines, J.F., V.F. McGowan, and V. BD. Skelman. 1986. World Directory of collections of micro-organisms. ed. 3, 678 p. World Data Center, Brisbane.
Trehane, P., C.D. Brickell, B.R. Baum, W.L.A. Hetterscheid, A.C. Leslie, J. McNeill, S.A. Spongberg, and F. Vrugtman, editors. 1995. The International Code of Nomenclature for Cultivated Plants — 1995. Quarterjack Publishing, Wimborne, U.K. p 175 (*Regnum veg.* 133).

Consultants

Werner Greuter
Botanischer Garten
Berlin, Germany

Noel H. Holmgren
New York Botanical Garden
Bronx, New York

David L. Hawksworth *
CAB-International Mycological Institute
Kew, England

---

* The authors of this section wish to express special thanks to Professor Hawksworth for his additions of material on fungal nomenclature, which were particularly helpful.

3

STATISTICS

Donald V. Sisson
Agricultural Experiment Station and
Department of Mathematics & Statistics
Utah State University
Logan, Utah 84322-4810

1. GENERAL TERMS

experimental unit That entity to which a given treatment is applied. Examples include a tree sprayed with a given chemical or a petri dish containing seed in a particular medium. In the latter example, the dish is the experimental unit, even if there are several seeds in the dish, and measurements are made on the individual seeds. The seeds are **samples** of the experimental unit.

experimental error (or MSE) Variability among experimental units that have been treated alike. Since many procedures assume equal variances within the treatments, the best estimate of experimental error involves combining or pooling the within-treatment variability. This estimate is usually called the **mean square error**, or simply the MSE (see *pooled variance* below).

replication The repeating of the application of a given treatment to more than one experimental unit. In the petri dish example of the definition of experimental unit, the seeds are not replications but are **samples**. These samples are sometimes referred to as **pseudoreplications**.

randomization The assignment of treatments to experimental units at random. This is done to obtain unbiased estimates of the treatment effects and mean square error. It removes personal bias or even the appearance of such bias.

local control (often called blocking) A restriction on the randomization imposed by the investigator in order to distribute systematic variability evenly among the treatments and to reduce the unexplained variability, or the MSE.

2. MEASURES OF CENTRAL TENDENCY

mean ($\bar{X}$) The arithmetic average of a set of values. This is the most efficient and common estimate of the "center" of a distribution but it is also affected the most by extreme values or outliers. It is usually denoted by

$$\bar{X} = \frac{\Sigma X}{n} \tag{1}$$

or the sum of all the observations (ΣX) divided by the number of observations *(n)*.

median The middle observation after the data have been ordered or ranked. If the number of observations is an even number, it is the average of the two middle numbers after ranking. It is not affected by outliers.

mode The observation that occurs with the greatest frequency. It is not very useful in small samples.

3. VARIABILITY

range *(R)* The distance between the largest and smallest observations.

$$R = X_{\text{maximum}} - X_{\text{minimum}}. \tag{2}$$

standard deviation *(S)* Approximately the average distance from the mean for a set of observations. It is usually denoted by

$$S = \sqrt{\frac{\Sigma(X - \bar{X})^2}{n-1}}\ . \tag{3}$$

If the data are normally distributed, or the distribution has the familiar bell-shaped curve, approximately two-thirds of the observations will be within one standard deviation of the mean and approximately 95% will be within two standard deviations of the mean.

variance *(S^2)* The square of the standard deviation (really just an intermediate step in the calculation of the standard deviation),

$$S^2 = \frac{\Sigma(X - \bar{X})^2}{n-1}\ . \tag{4}$$

coefficient of variation *(CV)* A measure of the relative variability when the standard deviation is expressed as a percentage of the mean and the units of measurement have been eliminated.

$$CV = \frac{S}{\bar{X}} \cdot 100 = \text{a percentage value.} \tag{5}$$

standard error of the mean ($S_{\bar{X}}$) Since the mean is itself a variable, it also has a standard deviation. This is denoted as

$$S_{\bar{X}} = \frac{S}{\sqrt{n}} \tag{6}$$

and is called the **standard error of the mean**. The standard error of the mean is to the mean what the standard deviation is to an individual observation.

standard error of the difference between two means ($S_{\bar{X}_1-\bar{X}_2}$) The difference between two means has a variance that is the sum of the variances of the individual means if the two means are independent. The standard error of this difference is the square root of the variance.

$$S_{\bar{X}_1-\bar{X}_2} = \sqrt{\frac{S_1^2}{n_1} + \frac{S_2^2}{n_2}} \,. \tag{7}$$

pooled variance If the assumption of equal variances holds, the information within groups is pooled to obtain

$$S_p^2 = \frac{(m_1-1)\,S_1^2 + (m_2-1)\,S_2^2}{(n_1-1) + (n_2-1)} \tag{8}$$

and the formula for standard error of the difference between two means becomes

$$S_{\bar{X}_1-\bar{X}_2} = \sqrt{S_p^2\left(\frac{1}{n_1} + \frac{1}{n_2}\right)} \,. \tag{9}$$

Note that S_p^2 is the same as **MSE** (mean square error) and can be expanded to accomodate any number of groups.

4. CONFIDENCE INTERVALS

A **confidence interval** is an interval estimate constructed in such a way that if a sampling experiment is repeated a large number of times and an interval constructed for each one, on the average a specified percentage of intervals will contain the true population value. If we choose a 95 % confidence level, we usually say that we are 95 % confident that our interval contains the true population value.

For the population mean, a confidence interval is found as follows:

$$\bar{X} \pm tS_{\bar{X}} \tag{10}$$

where t is a value from the table containing Student's t values (in almost all statistics books) corresponding to the confidence level desired and the degrees of freedom = n - 1.

For the population variance (σ^2), a confidence interval is found as follows:

$$\frac{(n-1)S^2}{\chi_1^2} \le \sigma^2 \le \frac{(n-1)S^2}{\chi_2^2} \tag{11}$$

where the χ^2 (chi-squared) values come from a table (found in most statistics books) corresponding to the appropriate confidence level and with degrees of freedom = n - 1.

For a proportion, a confidence interval is found as follows:

$$p \pm Z\sqrt{\frac{p(1-p)}{n}} \tag{12}$$

where

$$p = \frac{\#\ successes}{\#\ trials} \tag{13}$$

and Z is the standard normal variate, or t with degrees of freedom = ∞.

5. TEST OF HYPOTHESIS

A **hypothesis test** is a procedure to determine whether a proposed condition (hypothesis) is reasonable or not:

A. For a Population Mean. For a population mean, μ (mu), the condition is stated as μ_o where μ_o is a given value. For example, we could hypothesize that the true average mass of a set of samples was 6.5 g, or μ_o = 6.5 g. We use

$$t = \frac{\bar{X} - \mu_o}{S_{\bar{X}}}. \tag{14}$$

The hypothesis is rejected if the calculated t value exceeds the value in the t-table, with n - 1 degrees of freedom.

B. Difference Between Two Population Means. For the difference between two population means, the condition is stated as

$$\mu_1 - \mu_2 = \delta \tag{15}$$

where δ (delta) is a given value (usually 0) and

$$t = \frac{(\bar{X}_1 - \bar{X}_2) - \delta}{S_{\bar{X}_1 - \bar{X}_2}} \tag{16}$$

where t has $n_1 + n_2 - 2$ degrees of freedom if the population variances can be assumed to be equal.

C. Population Variance. For a population variance, the condition is stated as

$$\sigma^2 = \sigma_o^2 \quad \text{(sigma-squared)} \tag{17}$$

where σ_o^2 is a given value, and

$$\chi^2 = \frac{(n-1)S^2}{\sigma^2} \tag{18}$$

where χ^2 has n - 1 degrees of freedom and is compared to a table of χ^2 values, found in most statistics books.

D. Two Variances. For two variances, the condition is stated as $\sigma_1^2 = \sigma_2^2$, and

$$F = \frac{S_1^2}{S_2^2} \tag{19}$$

where F is compared to a table of F values found in most statistics books with $n_1 - 1$ and $n_2 - 1$ degrees of freedom for the numerator and denominator, respectively.

E. Population Proportion. For a population proportion, π (pi), the condition is stated as $\pi = \pi_o$, where π_o is a given value, and

$$Z = \frac{p - \pi_o}{\sqrt{\dfrac{\pi_o(1-\pi_o)}{n}}} . \tag{20}$$

F. Difference Between Two Proportions. For the difference between two proportions, the condition is stated as $\pi_1 - \pi_2 = \delta$, where δ is a given value, and

$$Z = \frac{(p_1 - p_2) - \delta}{\sqrt{\dfrac{p_1(1-p_1)}{n_1} + \dfrac{p_2(1-p_2)}{n_2}}} . \tag{21}$$

6. REGRESSION ANALYSIS

simple linear regression A procedure for relating two continuous variables when one variable (dependent variable) is expressed as a linear function of the other (independent variable). A common use is to predict one variable based on the information provided by the other. The form of the equation is

$$\hat{Y} = a + bX \tag{22}$$

where $\hat{Y}$ represents the predicted value.

multiple regression A procedure for expressing one dependent variable as a function of two or more independent variables.

least squares techniques One of the mathematical methods of obtaining estimates of the terms in a regression equation. This method minimizes the sum of the squares of the deviation of the observed *Y* variable (dependent) from the value as predicted by the regression equation.

slope In the linear regression equation,

$$\hat{Y} = a + bX$$

b is the slope of the line. It represents the predicted average unit change in *Y* per unit change in *X*. It is estimated (if the least squares technique is used) by the formula:

$$b = \frac{\Sigma(X_i - \bar{X})(Y_i - \bar{Y})}{\Sigma(X_i - \bar{X})^2} . \tag{23}$$

intercept In the linear regression equation,

$$\hat{Y} = a + bX$$

a is the Y intercept, or the predicted value of Y when $X = 0$. This may have no practical meaning in many problems, but it is still a necessary part of the equation. It is estimated (if the least squares technique is used) by the formula:

$$a = \bar{Y} - b\bar{X} \; . \tag{24}$$

standard error of estimate The square root of the residual (or unexplained) variance in a regression model. The formula is:

$$S_E = \sqrt{\frac{\Sigma\left(Y_i - \hat{Y}_i\right)^2}{n - 2}} \; . \tag{25}$$

standard error of the slope A measure of the variability of the slope of the regression line. It has the same relationship to the slope as the standard deviation has to the original variable, X. The formula is:

$$S_b = \frac{S_E}{\sqrt{\Sigma(X_i - \bar{X})^2}} \; . \tag{26}$$

correlation coefficient A measure of the mutual linear association between two continuous variables. It is an index as to how closely the actual points come to the predicted points. Perfect correlation is 1 (if the slope is positive) or -1 (if the slope is negative) and no correlation is represented by 0. The formula is:

$$r = \frac{\Sigma\left(X_i - \bar{X}\right)\left(Y_i - \bar{Y}\right)}{\sqrt{\Sigma\left(X_i - \bar{X}\right)^2 \, \Sigma\left(Y_i - \bar{Y}\right)^2}} \; . \tag{27}$$

coefficient of determination This represents the proportion of the variability in Y (dependent variable) that is predicted by X (independent variable). It is the square of the correlation coefficient.

7. ANALYSIS OF VARIANCE

Analysis of variance is a procedure for testing the equality of the means of two or more treatments by partitioning the variability into the amount caused by differences among the treatment means and the amount caused by differences among the experimental units within the treatments.

A. Experimental Designs. Experimental designs are the manner in which the treatments are assigned to the experimental units. Three most commonly encountered designs are:

i. Completely Randomized Design (CRD). The treatments are assigned to experimental units with no restrictions imposed. The linear model is

$$Y_{ij} = \mu + \tau_i + \epsilon_{ij} \qquad \textbf{(28)}$$

where Y_{ij} represents an individual experimental unit response,
μ is the overall mean,
τ_i (tau) is the effect of the ith treatment,
ϵ_{ij} (epsilon) is the random effect associated with the jth experimental unit assigned to the ith treatment.

If we are to assume there are k treatments with n experimental units in each group, and let $\bar{Y}_i$ be the average of all observations collected from experimental units assigned to i^{th} treatment and $\bar{Y}$ be the average of all the observations, the calculations for the analysis-of-variance table are in Table 1.

Table 1. Analysis-of-variance table for the completely randomized design.

Sources of Variation SV	Degrees of Freedom DF	Sums of Squares SS	Mean Square MS	F
Treatments	$k - 1$	$SST = n\sum_{i}^{k}(\bar{Y}_i - \bar{Y})^2$	$MST = \frac{SST}{k-1}$	$\frac{MST}{MSE}$
Experimental Error	$k(n - 1)$	$SSE = \sum_{i}^{k}\sum_{j}^{n}(Y_{ij} - \bar{Y}_i)^2$	$MSE = \frac{SSE}{k(n-1)}$	
Total	$kn - 1$	$\sum_{i}^{k}\sum_{j}^{n}(Y_{ij} - \bar{Y})^2$		

ii. Randomized Block Design (RBD). The experimental units are grouped or blocked in such a way that the variability from block to block is greater than the variability within blocks. Each treatment occurs once in each block. The linear model is

$$Y_{ij} = \mu + \tau_i + \beta_j + \epsilon_{ij} \qquad \textbf{(29)}$$

where Y_{ij} represents an individual experimental unit response,
μ is the overall mean
τ_i is the effect of the ith treatment
β_j (Beta) is the effect of the jth block.
ϵ_{ij} is the random effect associated with the experimental unit assigned to the ith treatment and occurring in the jth block.

If we assume there are k treatments arranged in b blocks and let $\bar{Y}_i$ be the average of the ith group, $\bar{Y}_j$ be the average of the jth block, and $\bar{Y}$ be the overall average, the calculations for the analysis-of-variance table are in Table 2.

Table 2. Analysis-of-variance table for the randomized block design.

SV	DF	SS	MS	F
Treatments	$k-1$	$SST = b\sum_{i}^{k}(\bar{Y}_i - \bar{Y})^2$	$MST = \frac{SST}{k-1}$	$\frac{MST}{MSE}$
Blocks	$b-1$	$SSB = k\sum_{j}^{b}(\bar{Y}_j - \bar{Y})^2$	$MSB = \frac{SSB}{b-1}$	
Experimental Error	$(k-1)(b-1)$	$SSE = \sum_{i}^{k}\sum_{j}^{b}(Y_{ij} - \bar{Y}_i - \bar{Y}_j + \bar{Y})^2$	$MSE = \frac{SSE}{(k-1)(b-1)}$	
Total	$kb-1$	$\sum_{i}^{k}\sum_{j}^{b}(Y_{ij} - \bar{Y})^2$		

iii. Latin square design. In the latin square design the experimental units are grouped or blocked in two dimensions (usually designated as rows and columns) as opposed to one dimensional blocking in the randomized block design. Each treatment occurs once in each row and once in each column. The number of treatments is equal to the number of rows and the number of columns. The linear model is

$$Y_{ijk} = \mu + \tau_i + \beta_j + \gamma_k + \epsilon_{ijk} \qquad \textbf{(30)}$$

where Y_{ijk} represents an individual experimental unit response,
μ is the overall mean,
τ_i is the effect of the ith treatment,
β_j is the effect of the jth row,
γ_k (gamma) is the effect of the kth column,
ϵ_{ijk} is the random effect associated with the experimental unit in the jth row and k^{th} column that was assigned to the ith treatment.

If we let k equal the number of treatments (or rows or columns) and $\bar{Y}_i$ be the average of the ith treatment, $\bar{Y}_j$ be the average of the jth row, $\bar{Y}_k$ be the average of the kth column, the calculations for the analysis-of-variance table are in Table 3.

Table 3. Analysis of variance table for the latin square design.

SV	DF	SS	MS	F
Treatments	$k - 1$	$SST = k\sum_{i}^{k}(\bar{Y}_i - \bar{Y})^2$	$MST = \frac{SST}{k-1}$	$\frac{MST}{MSE}$
rows	$k - 1$	$SSR = k\sum_{j}^{k}(\bar{Y}_j - \bar{Y})^2$	$MSR = \frac{SSR}{k-1}$	
columns	$k - 1$	$SSC = k\sum_{k}^{k}(\bar{Y}_k - \bar{Y})^2$	$MSC = \frac{SSC}{k-1}$	
Experimental Error	$(k-1)(k-2)$	$SSE = \sum_{i}^{k}\sum_{j}^{k}(Y_{ij} - \bar{Y}_i - \bar{Y}_j - \bar{Y}_k + 2\bar{Y})^2$	$MSE = \frac{SSE}{(k-1)(k-2)}$	
total	$k^2 - 1$	$\sum_{i}^{k}\sum_{j}^{k}(Y_{ij} - \bar{Y})^2$		

B. Mean Comparisons. Mean comparisons are procedures employed where some or all of the k treatments means (or averages) are compared in an attempt to interpret the results of the F-test in an analysis of variance.

i. Planned comparisons. Planned comparisons are specific comparisons that are of obvious interest, even before the experiment is conducted.

a) Factorial experiments. Factorial experiments are experiments in which the treatments consist of all possible combinations of the different levels of two or more factors studied simultaneously. As an example, consider the response of a plant to conditions when both temperature and humidity are varied. Let temperature be factor A with a = 3 levels and humidity be factor B with b = 2 levels. The resulting experiment would have 3 x 2 = 6 treatments where a treatment denotes a particular combination of temperature and humidity. Assume that there are r experimental units in each treatment. Table 4 shows the schematic layout of means.

Table 4. Means in a two-way factorial experiment.

		A = temperature			
		a_1	a_2	a_3	Row Average
B = humidity	b_1	$\bar{Y}_{11.}$	$\bar{Y}_{21.}$	$\bar{Y}_{31.}$	$\bar{Y}_{.1.}$
	b_2	$\bar{Y}_{12.}$	$\bar{Y}_{22.}$	$\bar{Y}_{32.}$	$\bar{Y}_{.2.}$
Column Average		$\bar{Y}_{1..}$	$\bar{Y}_{2..}$	$\bar{Y}_{3..}$	$\bar{Y}_{...}$ = overall average

Main Effect is the effect of one factor averaging over the levels of all of the other factors. These are tested in the analysis of variance using an F test (where the mean square error is usually the denominator), as indicated in Table 5.

Table 5. Partial analysis of variance table for a two-way factorial experiment–main effects.

SV	df	SS	MS	F
A main effect	$a-1$	$SSA = rb\sum_{i}^{a}(\bar{Y}_{i..} - \bar{Y}_{...})^2$	$MSA = \frac{SSA}{a-1}$	$\frac{MSA}{MSE}$
B main effect	$b-1$	$SSB = ra\sum_{j}^{k}(\bar{Y}_{.j.} - \bar{Y}_{...})^2$	$MSB = \frac{SSB}{b-1}$	$\frac{MSB}{MSE}$

Interaction is the situation where differences among the levels of one factor, say factor A, change from level to level of the second factor, say factor B. The test for an interaction is also made in the analysis of variance table as indicated in Table 6.

Table 6. Partial analysis of variance table for a two-way factorial experiment–interaction.

SV	df	SS	MS	F
AB interaction	$(a-1)(b-1)$	$SSAB = r\sum_{i}^{a}\sum_{j}^{b}(\bar{Y}_{ij.} - \bar{Y}_{i..} - \bar{Y}_{.j.} + \bar{Y}_{...})^2$	$MSAB = \frac{SSAB}{(a-1)(b-1)}$	$\frac{MSAB}{MSE}$

Simple effect is the effect of one factor when all other factors are held constant. These are not tested directly in the analysis of variance table. The simple effect of B at the a_1 level would be estimated by $\bar{Y}_{12.} - \bar{Y}_{11.}$. It could be tested by using the concept of linear comparisons.

b) Linear comparisons. Linear comparisons are contrasts between any 2 sets (one or more means in each set) of means. The simple effect illustrated above is an example, and the test would be:

$$t = \frac{\bar{Y}_{12.} - \bar{Y}_{11.}}{\sqrt{\frac{2MSE}{r}}} \tag{31}$$

In general, if the linear comparison is of the form $a\bar{Y}_1 \pm b\bar{Y}_2 \pm c\bar{Y}_3$, etc, the variance of the linear combination is given by

$$a^2 V(\bar{Y}_1) + b^2 V(\bar{Y}_2) + c^2 V(\bar{Y}_3) = a^2\left(\frac{MSE}{n_1}\right) + b^2\left(\frac{MSE}{n_2}\right) + c^2\left(\frac{MSE}{n_3}\right) \tag{32}$$

(This assumes that the $\bar{Y}_i$'s are independent.)

ii. All possible comparisons.

Fisher's least significant difference test (LSD). All possible differences among the means are compared with the LSD value when the

$$LSD = t\sqrt{\frac{2MSE}{m}} \tag{33}$$

where the degrees of freedom for the *t* are the degrees of freedom associated with the MSE. This test has a high Type I error rate (which also gives a low Type II error rate.)

Tukey's Test. The critical value is

$$q\sqrt{\frac{MSE}{m}} \tag{34}$$

when *q* is a value taken from a studentized range table (available in many statistical textbooks). The Type I error rate is low (hence the Type II error rate is high).

Newman-Keul's Test and Duncan's Test. Tests with intermediate (between the LSD and Tukey's Test) Type I error rates. These are accomplished by ranking the means to be compared and using different critical values for different ranges where two means adjacent in the rankings have a range of 2, one other mean between them gives a range of 3, etc.

iii. Orthogonal Polynomials. A comparison among means when regression effects are emphasized and the objective is to estimate the form of the response, such as linear, quadratic, cubic, etc. The calculations are similar to linear combinations, with appropriate weighting coefficients derived for each term in the polynomial.

C. Variance Components. Variability in a linear model is contributed by two or more effects. In the model

$$Y_{ij} = \mu + \tau_i + \epsilon_{ij}$$

the random variability associated with the ϵ_{ij} effect can be designated as σ^2_ϵ. Likewise, the variability introduced by the treatment effects, τ_i, can be designated as σ^2_τ if treatments are considered to be random. Both σ^2_E and σ^2_T are variance components of this model.

8. COVARIANCE ANALYSIS

Covariance Analysis is a combination of regression and analysis of variance that allows mean comparisons among treatments in the dependent variable to be made after adjusting for effects of the independent variable. In addition, the MSE is based on deviations from a regression model rather than deviations from the mean, hence the MSE is usually smaller, and we have a gain in precision.

As an example of the formulae involved, consider a randomized block design [see the model in equation (29)] when the amount of nitrogen produced by alfalfa plants is measured under different moisture-stress treatments. Each experimental unit consists of 25 seeds. Since germination rates may vary, the number germinating may be used as the independent variable X. The linear model is

$$Y_{ij} = \mu + \tau_i + \beta_j + \rho X_{ij} + \epsilon_{ij} \quad (35)$$

where the new term, ρX_{ij}, is the effect of the germination on that experimental unit. (See table 6.)

Table 6. Simple linear analysis of covariance table for a randomized block design.

SV	dF	SS_X	SP	SS_Y	Deviations from Regression		
					dF	SS	MS
treatments	$k-1$	$SST_x = b\sum_i^k(\bar{X}_i-\bar{X})^2$	$SPT = b\sum_i^k(\bar{X}_i-\bar{X})(\bar{Y}_i-\bar{Y})$	$SST_Y = b\sum_i^k(\bar{Y}_i-\bar{Y})^2$			
blocks	$b-1$	$SSB_x = k\sum_j^b(\bar{X}_j - \bar{X})^2$	$SPB = k\sum_j^b(\bar{X}_j-\bar{X})(\bar{Y}_j-\bar{Y})$	$SSB_Y = b\sum_j^b(\bar{Y}_j-\bar{Y})^2$			
experimental error	$(k-1)(b-1)$	$SSE_x = \sum_i^k\sum_j^b(X_{ij}-\bar{X}_i-\bar{X}_j+\bar{X})^2$	$SPE =$ $\sum_i^k\sum_j^b(X_{ij}-\bar{X}_i-\bar{X}_j+\bar{X})(Y_{ij}-\bar{Y}_i-\bar{Y}_j+\bar{Y})$	$SSE_Y = \sum_i^k\sum_j^b(Y_{ij}-\bar{Y}_i-\bar{Y}_j+\bar{Y})^2$	$(k-1)(b-1)-1$	$SSE = SSE_Y - \frac{(SPE)^2}{SSE_X}$	$\frac{SSE}{(k-1)(b-1)-1}$
treatment plus experimental error	$b(k-1)$	SST_X+SSE_x	$SPT+SPE$	SST_Y+SSE_Y	$b(k-1)-1$	$SS(T+E) =$ $SST_Y+SSE_Y-\frac{(SPT+SPE)^2}{SST_X+SSE_X}$	
(treatment plus experimental error) - error = Adj. Means					$k-1$	$SS_{Adj\ Means} =$ $SS(T+E)-SSE$	$\frac{SS_{Adj\ Means}}{k-1}$

F for testing the equality of adjusted means = $\frac{MS_{Adj\ Means}}{MSE}$

9. NONPARAMETRIC TESTS

These are tests that make a few or no assumptions regarding the underlying distribution of the variable. The power is usually less than that of a corresponding parametric test.

A. Sign Test. A test for the median of a population. It classifies each observation as being either above (+) or below (-) the hypothesized median and then tests to see if the observed proportion above the median, P, differs from 0.5 by using either standard binomial tables or the normal approximation to the binomial:

$$Z = \frac{P - 0.5}{\sqrt{\frac{(0.5)(1-0.5)}{n}}} \tag{36}$$

B. Wilcoxon's Signed-Rank Test. A paired comparison test where the absolute differences between pair members are ranked, then reassigned their original sign. If there is no difference between the two groups, the expected value of the sums of the signed ranks should be 0. As in many of the non-parametric tests, a special table is used to see if the difference is significant.

C. Mann-Whitney Two-Sample Test. A test for the equality of two population means where the data for both groups are pooled and ranked. Each ranking is then assigned its accompanying group identification. The sum of the group with the smaller sample size, R, is obtained. The test is a Z score of the form

$$Z = \frac{R - \mu_R}{\sigma_R} = \frac{R - \frac{n_1(n_1 + n_2 + 1)}{2}}{\sqrt{\frac{n_1 n_2 (n_1 + n_2 + 1)}{12}}} \tag{37}$$

D. Kruskal-Wallis *k*-Sample Test. A test for the equality of the means of k different samples. It is the counterpart of the Analysis of Variance. All of the data from the k groups are ranked as one combined sample, and the group identification is then reassigned to each rank value. The sums of the ranks, R_i, are then obtained, and a chi-squared test is performed as follows:

$$\chi^2 = \frac{12}{\sum_i^k n_i \left(\sum_i^k n_i + 1 \right)} \cdot \sum_i^k \frac{R_i^2}{n_i} - 3\left(\sum_i^k n_i + 1 \right) \tag{38}$$

with k - 1 degrees of freedom.

E. Contingency Tables. Count tables where the experimental (or survey) units are classified according to two or more discrete variables in an attempt to determine whether the variables are related or independent. There are many techniques for analyzing these tables (categorical data analysis), but for two factors illustrated in Table 7, a chi-squared test is made as follows:

$$\chi^2 = \sum_i^r \sum_j^c \frac{(O_{ij} - E_{ij})^2}{E_{ij}} \tag{39}$$

with $df = (r - 1)(c - 1)$

where $E_{ij} = \frac{R_i C_j}{n}$, or the expected number in the ith row and j^{th} column. **(40)**

r = the number of rows,

and c = the number of columns.

Table 7. Two-way contingency table.

	Germinating	Not Germinating	R_i [b]
Group 1	O_{11} [a]	O_{12}	R_1
Group 2	O_{21}	O_{22}	R_2
Group 3	O_{31}	O_{32}	R_3
C_j [c]	C_1	C_2	n [d]

[a] O_{ij} represents the number of individuals in the ith row and j^{th} column.
[b] R_i represents the total of the ith row.
[c] C_j represents the total of the jth column.
[d] n is the total sample size.

10. MISCELLANEOUS

A. Central Limit Theorem. One of the most important practical theorems in statistics. It basically says that as the sample size increases, the distribution of the sample mean will be normal with a mean of μ and a standard error of $\frac{\sigma}{\sqrt{n}}$.

B. Sample Size. The number of experimental units used in each treatment of an experiment. An approximation to the number required is

$$n = \frac{Z^2 \sigma^2}{D^2} \tag{41}$$

where n is the required sample size,
Z is the standard normal variable (1.96 if one is working with an α-level of 0.05).
σ is the population standard deviation,
and D is the size of the effect one wishes to detect as a "significant" effect.

Consultants

Paul N. Hinz
Iowa State University
Ames, Iowa

Gary Richardson
Colorado State University
Fort Collins, Colorado

II

PLANT BIOPHYSICS

The science of plant physiology relies heavily upon a variety of biophysical measurements. It is the goal of this section to summarize the symbols, units, and terms that are used to express the results of these measurements. Uniformity of expression seems highly desirable, so this section emphasizes recommended SI units and symbols. At the same time, it is recognized that many plant scientists will continue to use other units and symbols, so acceptable alternatives are presented in a few cases (e.g., millimoles per liter as an alternative to moles per cubic meter). Definitions of biophysical terms are also given.

Many physical parameters can be considered as pairs, with one of the pair expressing a quantity and its partner expressing a potential for transfer of the quantity across a barrier. Thus, joules per kilogram are used as units to express the quantity of heat energy in some substance under consideration, while temperature differences are used to express the potential for transfer of heat from one point to another (i.e., from a point of higher temperature to a point of lower temperature). In a thermodynamic system, the parameter expressing the quantity is said to be **extensive** (the value is the sum of the value for subdivisions of the system; e.g., volume), and the parameter expressing the transfer potential is said to be **intensive** (has the same value for any subdivision of the system; e.g., pressure). Mass factors lend themselves well to such an analysis. The quantity of mass is expressed as kilograms or moles, while the potentials for transfer are expressed in various ways:

gases	pressure or partial pressure (pascals)
water	water potential (pascals, joules per kilogram)
hydrogen ions in water	*p*H units
solutes in water	chemical potential (or concentration or negative logarithm of concentration or activity)

Biophysical measurements in plant physiology are ultimately dependent upon the concepts developed in physics and chemistry. To a great extent, they depend upon thermodynamics; hence, that is the first topic of this section.

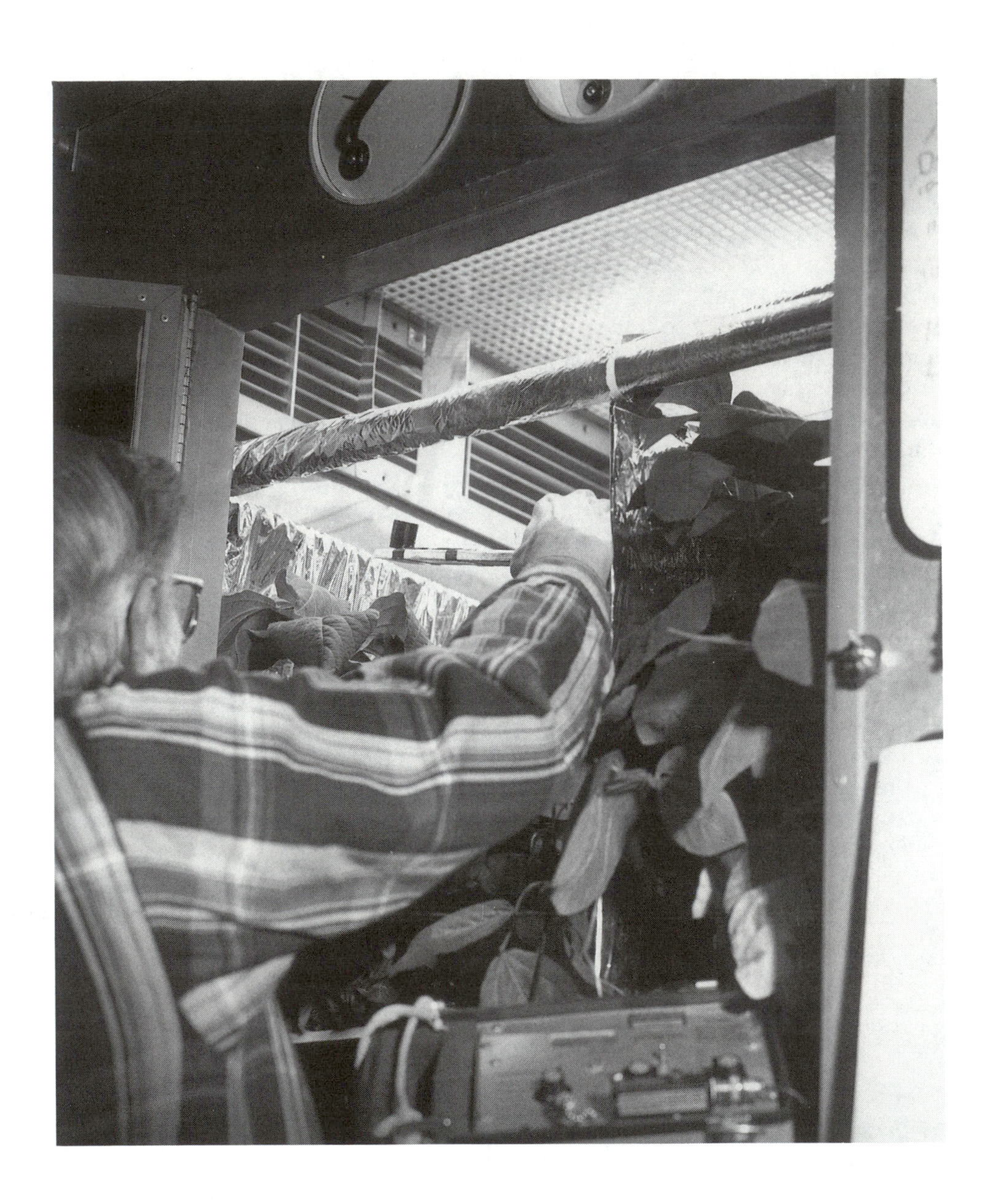

4

BASIC THERMODYNAMIC QUANTITIES

Michael J. Savage
Department of Agronomy
University of Natal
Pietermaritzburg 3201
Republic of South Africa

The purpose of this chapter is to present a simple treatment of some of the basic thermodynamic concepts involved in plant physiology, and especially those relating to water potential and its measurement. The concepts will be briefly applied to plant water potential and to water potential measurement techniques.

1. BASIC CONCEPTS AND THE CHEMICAL POTENTIAL

Purely from an energy conservation standpoint, one would expect that, for a closed system (that is, one with constant mass), the total (internal) energy change of the system (dE, joules) is the heat energy (dQ) added to the system minus the work done (dW) by the system:

$$dE = dQ - dW. \quad (1)$$

This expression is commonly referred to as the **first law of thermodynamics**. The internal energy E represents the kinetic and potential energies of the molecules, atoms, and subatomic particles that constitute the (closed) system on a macroscopic scale; there is presently no known way to determine E absolutely. Only changes in E, dE, are required, however, and these changes are normally determined by experiment. It is also important to recognize that the energy of the system is totally conserved and that the energy available for useful work is continually decreasing and being converted into energy unavailable for useful work. Indeed, it is this unavailable energy that provides insight into the concept of system entropy.

In simple terms, the change of entropy of a system, dS (unit J K^{-1}), is the ratio of the heat energy added to the system (dQ) to the thermodynamic temperature T of the system and, in accordance with thermodynamic principles, is always increasing for any real system. These two statements are often expressed mathematically in differential form as:

$$dS = dQ/T \tag{2}$$

and

$$dS \geq 0. \tag{3}$$

Entropy was first introduced in classical thermodynamics to provide a quantitative basis for the common observation that naturally occurring processes have a particular direction. For example, the flow of heat energy occurs from a hotter to a cooler region. Equations 2 and 3 represent statements of the **second law of thermodynamics.**

Mechanical work occurs when work done on a system results in motion. If F (with unit N) is the component of the **force** acting in the direction of the displacement (dl), the mechanical work, dW, equals $F \cdot dl$. In this case, $dW = F \cdot dl = (F/A) \cdot (A \cdot dl) = P \cdot dV$, where P, which is force (F)/area (A), is the external pressure exerted on the system resulting in a volume change, dV $(= A \cdot dl)$.

Combining the first and second laws of thermodynamics (Equations 1 and 2) and the equation $dW = P \cdot dV$ describing the mechanical work done by the system, we have:

$$dE = T \cdot dS - P \cdot dV. \tag{4}$$

This equation describes the internal energy change for reversible and purely mechanical thermodynamic processes. However, different systems may be subjected to work done by a number of forces that may or may not include pressure (i.e., mechanical) work forces. Other work forces could include work of magnetization, electrical work, etc. To allow for the possibility of other work forces involved in the closed system currently considered, we write:

$$dE = T \cdot dS - P \cdot dV - dW \tag{5}$$

where dW now represents the total of all other forms of work done by the system on the surroundings. This work term can be expressed as:

$$dW = -\sum_j Y_j \cdot dX_j \tag{6}$$

where the jth work term is the product of an intensive parameter Y_j and an extensive parameter X_j (Babcock, 1963; Bolt and Frissel, 1960). (See Chapter 5 for list of subscripts.)

Including all forms of work done by the system on the surroundings, the change in internal energy for a closed system is expressed by:

$$dE = T \cdot dS - P \cdot dV + \sum_j Y_j \cdot dX_j. \tag{7}$$

In chemical thermodynamics, it is common for the composition of the system to be varied; i.e., not closed. In such an open system, in addition to the variables of entropy (S), volume (V), and extensive parameter X_j (for given corresponding T, P, and Y_j), the composition of the system is varied. The amount of substance of chemical species component i, n_i (with unit mol) is used to describe the chemical composition of the system. Then, choosing S, V, X_j, and n_i as independent system variables, we note that the change in E could be due to independent changes in S, V, X_j, and n_i. The change in E, for example, due to change in S only, with V, X_j, and n_i held constant, can be expressed mathematically as:

$$\left(\frac{\partial E}{\partial S}\right)_{V,X_j,n_i} \cdot \mathrm{d}S .$$

The term $\partial E / \partial S$ is called a partial derivative because it expresses the change in E with respect to S only.

Hence, dE = (change in E with respect to S only) + (change in E with respect to V only) + $\sum_j$ (change in E with respect to X_j only) + $\sum_i$ (change in E with respect to n_i only).

Hence:

$$\mathrm{d}E = \left(\frac{\partial E}{\partial S}\right)_{V,X_j,n_i} \cdot \mathrm{d}S + \left(\frac{\partial E}{\partial V}\right)_{S,X_j,n_i} \cdot \mathrm{d}V + \sum_j \left(\frac{\partial E}{\partial X_j}\right)_{V,S,n_i} \cdot \mathrm{d}X_j + \sum_i \left(\frac{\partial E}{\partial n_i}\right)_{S,V,X_j} \cdot \mathrm{d}n_i . \tag{8}$$

We define the **chemical potential** μ_i (J mol^{-1}) of the ith solute species by the partial change in internal energy E with respect to n_i, the amount of substance of chemical species component i, with entropy S, volume V, extensive parameter X_j, and other solute species $n_k(k \neq i)$ kept constant:

$$\mu_i = \left(\frac{\partial E}{\partial n_i}\right)_{S,V,X_j,n_k(k \neq i)} . \tag{9}$$

From Equation 8 and incorporating the definition of chemical potential of the ith solute species, μ_i (Equation 9), we get:

$$\mathrm{d}E = T \cdot \mathrm{d}S - P \cdot \mathrm{d}V + \sum_j Y_j \cdot \mathrm{d}X_j + \sum_i \mu_i \cdot \mathrm{d}n_i . \tag{10}$$

2. FREE ENERGY AND WATER POTENTIAL

The **Gibbs free energy** G (J) is defined by:

$$G = E + PV - TS. \tag{11}$$

The Gibbs free energy represents the energy available for useful work. Hence, the difference in the Gibbs free energy between two states can be used to predict the spontaneous direction for a process and indicates the useful work the transition makes available. From Equation 11, we get:

$$\mathrm{d}G = \mathrm{d}E + P \cdot \mathrm{d}V + V \cdot \mathrm{d}P - T \cdot \mathrm{d}S - S \cdot \mathrm{d}T \tag{12a}$$

and hence, from Equation 10,

$$dG = -S \cdot dT + \sum_j Y_j \cdot dX_j + \sum_i \mu_i \cdot dn_i \tag{12b}$$

so that:

$$\left(\frac{\partial G}{\partial n_i}\right)_{T,P,X_j,n_k (k \neq i)} = \mu_i. \tag{13}$$

This equation expresses the relationship between the Gibbs free energy and the chemical potential of species *i*. The chemical potential of species *i* roughly indicates the free energy associated with it and available for performing work. For instance, considering the chemical potential for water, μ_w,

$$\mu_w = \left(\frac{\partial G}{\partial n_w}\right)_{T,P,X_j,n_k} \tag{14}$$

where n_k cannot be n_w, then:

$$dG = -S \cdot dT + V \cdot dP + \sum_j Y_j \cdot dX_j + \mu_w \cdot dn_w + \sum_{i'} \mu_i \cdot dn_{i'}$$

where the i' indicates that the summation cannot include the water component as it has already been included. Choosing T, P, X_j and n_k as independent variables for μ_i, it can be shown that, where ℓ is a dummy variable in that ℓ would not appear if the summation were written out:

$$d\mu_i = -\bar{S}_i \cdot dT + \bar{V}_i \cdot dP + \sum_j \bar{Y}_j \cdot dX_j + \sum_k \left(\frac{\partial \mu_i}{\partial n_k}\right)_{T,P,X_j,n_\ell (\ell \neq k, k \neq i)} \cdot dn_k$$

where

$$\left(\frac{\partial S}{\partial n_i}\right)_{T,P,X_j,n_k (k \neq i)} = \bar{S}_i \qquad (\text{J mol}^{-1}),$$

$$\left(\frac{\partial V}{\partial n_i}\right)_{T,P,X_j,n_k (k \neq i)} = \bar{V}_i \qquad (\text{J mol}^{-1}),$$

and

$$\left(\frac{\partial Y_j}{\partial n_i}\right)_{T,P,X_j,n_k (k \neq i)} = \bar{Y}_j \qquad (\text{J mol}^{-1}).$$

The quantities $\bar{S}_i$, $\bar{V}_i$, and $\bar{Y}_j$ are partial molar values for entropy, volume, and intensive parameter Y_j, respectively. Defining the ith chemical species to be water, we obtain

$$d\mu_w = d(\mu_w - \mu_w^*)$$

$$= -\bar{S}_w \cdot dT + \bar{V}_w \cdot dP + \sum_j \bar{Y}_j \cdot dX_j + \sum_k \left(\frac{\partial \mu_w}{\partial n_k}\right)_{T,P,X_j,n_\ell(\ell \neq k,\, k \neq w)} \cdot dn_k \tag{15}$$

where we define μ_w^* to be the chemical potential of pure free water at a pressure of 101.3 kPa and at the same temperature as the water with chemical potential μ_w. Under these isothermal conditions, the temperature difference indicated by dT is zero so that:

$$d(\mu_w - \mu_w^*) = \bar{V}_w \cdot dP + \sum_j \bar{Y}_j \cdot dX_j + \sum_k \left(\frac{\partial \mu_w}{\partial n_k}\right)_{T,P,X_j,n_\ell(\ell \neq k,\, k \neq w)} \cdot dn_k. \tag{16}$$

Following integration of Equation 16, we define:

$$\Psi_m = \frac{\mu_w - \mu_w^*}{\bar{M}_w}, \tag{17}$$

$$\Psi_v = \frac{\mu_w - \mu_w^*}{\bar{V}_w}, \tag{18}$$

$$\Psi_f = \frac{\mu_w - \mu_w^*}{\bar{M}_w g}, \tag{19}$$

and

$$\Psi_n = \mu_w - \mu_w^*, \tag{20}$$

where $\bar{M}_w$ (kg mol^{-1}) is the partial molar mass of water, the subscripts $_{m,\ v,\ f,}$ and $_n$ refer to unit mass, volume, weight (a force), and amount of substance, respectively; and g (m s^{-2}) is the acceleration due to gravity, $\bar{V}_w$ is the partial molar volume of water, and Ψ_m (J kg^{-1}), Ψ_v (J m^{-3}, N m^{-2} or Pa), Ψ_f (J N^{-1} or m), and Ψ_n (J mol^{-1}) refer to the specific, volumetric, weight, and molar water potentials, respectively (Rose, 1979; Savage, 1978).

Water potential is the amount of useful work per unit quantity of water done by means of externally applied forces in transferring, reversibly and isothermally, an infinitesimal amount of water from some standard reference state to its position in the soil, plant, or atmosphere. The reference state is that of pure free water at the same temperature as the water in the system and at a pressure of one standard atmosphere, namely, 101.3 kPa (adapted from Taylor and Ashcroft, 1972, p 153 and Bolt et al., 1975). The SI unit of work is the joule (J).

Water potential may be expressed as the amount of useful work per unit mass, volume, weight, or amount of substance (mol) of water. Plant physiologists use the symbol Ψ for water potential and usually define it to correspond to a volume basis (Ψ_v). Some workers have used a mass basis (Ψ_m) and others an amount-of-substance basis (Ψ_n). In any system of units,

$$\Psi = \rho_w \, \Psi_m = \Psi_v = \rho_w g \, \Psi_f = \frac{\Psi_n}{\bar{V}_w}$$

where, using SI units, ρ_w (kg m^{-3}) is the density of liquid water, where $\rho_w = \rho_w(T)$, T(°C) is the water temperature, and $\rho_w = \bar{M}_w / \bar{V}_w$.

3. ENTHALPY

The enthalpy (H) of a system is defined as:

$$H = E + PV. \tag{22}$$

As in the case of the Gibbs free energy, consider the change of the function, in this case the enthalpy, from an initial equilibrium state to a final equilibrium state. Therefore,

$$\begin{aligned} \mathrm{d}H &= \mathrm{d}E + P \cdot \mathrm{d}V + V \cdot \mathrm{d}P \\ &= \mathrm{d}Q + V \cdot \mathrm{d}P \end{aligned}$$

where

$$\mathrm{d}E = \mathrm{d}Q - P \cdot \mathrm{d}V.$$

Hence, for an isobaric process, dH is equal to dQ, the heat energy amount transferred. In thermodynamic chemistry where isobaric processes are more important than isovolumic processes, enthalpy is of greatest use. From the definition of Gibbs free energy, the change in enthalpy can be defined *via*:

$$\mathrm{d}G = \mathrm{d}H - T \cdot \mathrm{d}S \quad \text{or} \quad \mathrm{d}H = \mathrm{d}G + T \cdot \mathrm{d}S. \tag{23}$$

4. WATER POTENTIAL IN THE VAPOR STATE

Consider water vapor (which may be just one component of the gas phase) behaving as an ideal gas. Then,

$$e\bar{V}_w = RT \tag{24}$$

where e (Pa) is the partial water vapor pressure, and R (8.3143 J mol^{-1} K^{-1}) is the Universal gas constant. We have, from Equation 16,

$$\frac{\partial}{\partial P}(\mu_w - \mu_w^*) = \bar{V}_w. \tag{25}$$

Integrating this equation over a vapor pressure range from $P = e_o$ (the saturation vapor pressure) to $P = e$ and applying Equation 24:

$$\mu_w - \mu_w^* = \int_{e_o}^{e} \left(\frac{RT}{P}\right) dP$$

$$= RT \ln\left(\frac{e}{e_o}\right).$$

It has been assumed that isothermal conditions prevail during the change in pressure from e_o to e and that $e > 0$ kPa.

Substitution of Equation 18 into the equation above yields the Kelvin equation expressing water potential $\Psi = \Psi_v$ (Pa) as a function of fractional relative humidity e/e_o:

$$\Psi = \Psi_v = \left(\frac{RT}{\bar{V}_w}\right) \ln\left(\frac{e}{e_o}\right). \qquad (26)$$

5. COMPONENTS OF WATER POTENTIAL

Three terms emerge from the thermodynamic theory as being components of water potential $d\Psi_n$ (J mol^{-1}). In differential form, from Equation 15 applied to isothermal conditions, we have:

$$d\Psi = \bar{V}_w \cdot dP + \sum_j Y_j \cdot dX_j + \sum_k \left(\frac{\partial \mu_w}{\partial n_k}\right)_{T,P,X_j,n_{\ell(\ell \neq k,\, k \neq w)}} \cdot dn_k. \qquad (27)$$

It is desirable to partition water potential into components even if there is some doubt about the partitioning (Spanner, 1973). Ignoring the work term, the second term of Equation 27, and writing $\bar{V}_w = \partial\mu_w / \partial P$ and separating

$$\sum_k \left(\frac{\partial \mu_w}{\partial n_k}\right)_{T,P,X_j,n_{\ell(\ell \neq k,\, k \neq w)}} \cdot dn_k$$

into a water part and a non-water part, we have:

$$d\mu_w = \left(\frac{\partial \mu_w}{\partial P}\right) \cdot dP + \sum_k \left(\frac{\partial \mu_w}{\partial n_k}\right)_{T,P,X_j,n_{\ell(\ell \neq k,\, k \neq w)}} \cdot dn_k + \left(\frac{\partial \mu_w}{\partial n_w}\right)_{T,P,X_j,n_{\ell(\ell \neq k,\, k \neq w)}} \cdot dn_w. \qquad (28)$$

The first term, which can be positive or negative, can be written as $\bar{V}_w \cdot dP$ and expresses the dependence of the chemical potential of water, μ_w, on pressure P. In the older literature, this term is written as dP but in more modern literature the symbol $d\Psi_p$ is used. Generally, Ψ_p is termed the **pressure potential.**

The second term of the right hand side of Equation 28,

$$\sum_k \left(\frac{\partial \mu_w}{\partial n_k}\right)_{T,P,X_j,n_{l\,(l \neq k,\, k \neq w)}} \cdot dn_k,$$

arises from the contribution of the dissolved solutes to the chemical potential of the water (Dainty, 1976; Slatyer, 1967), commonly referred to as the osmotic component, and may be written in traditional notation as $-\bar{V}_w \cdot d\pi_k$ where π_k (Pa) is referred to as the osmotic pressure arising from the kth component. In more modern literature, the second term is written as $\bar{V}_w \cdot d\Psi_\pi$ or sometimes $\bar{V}_w \cdot d\Psi_s$ where Ψ_π (Pa) is termed the **osmotic potential** and Ψ_s (Pa) the **solute potential.** The old (or traditional) term osmotic pressure, π_k, is always positive whereas the more recent term osmotic potential Ψ_π (or the solute potential, Ψ_s), is always negative.

The third term of Equation 28,

$$\left(\frac{\partial \mu_w}{\partial n_w}\right)_{T,P,X_j,n_{k(l \neq k,\, k \neq w)}} \cdot dn_k$$

expresses the matric component arising from the solid matter in the system, in which the chemical potential of water is a function of water content (Dainty, 1976; Slatyer, 1967), may be written in traditional notation as $-\bar{V}_w \cdot d\tau$ where $-\tau$ (Pa) is referred to as the **matric potential.** Using the notation of the more modern literature, the third term of Equation 28 is usually written $\bar{V}_w \cdot d\Psi_m$ where Ψ_m (Pa) is also referred to as the matric potential. The old (or traditional) term τ is always positive whereas the more recent term Ψ_m is negative.

Integrating Equation 28 and substituting for the various water potential components, we have, with all terms having Pa as their unit:

$$\Psi = P - \pi - \tau \tag{29}$$

in traditional notation, or, in more modern potential terminology,

$$\Psi = \Psi_p + \Psi_s + \Psi_m. \tag{30}$$

Of particular note is the controversy regarding matric potential ($-\tau$ in Equation 29 and Ψ_m in Equation 30) as a component of the total water potential. Some workers (Passioura, 1980; Salisbury and Ross, 1991) doubt that matric potential Ψ_m can be included in Equation 28 as shown also in Equations 29 and 30.

6. WATER POTENTIAL OF AQUEOUS SOLUTIONS

Applying Equation 27, valid for isothermal conditions only, under conditions of constant pressure and in the absence of any work fields, water potential becomes a function of composition and concentration only. Combining Equations 26 and 27 and considering a solution containing only one solute, say NaCl, we get:

$$d\Psi = \left(\frac{1}{\bar{V}_w}\right)\left(\frac{\partial \mu_w}{\partial n_{NaCl}}\right)_{T,P,X_j} \tag{31}$$

where Ψ (Pa) is related to vapor pressure (Equation 26).

Equation 31 forms the basis for the use of thermocouple hygrometers. Essentially these are instruments containing solutes, liquid, and vapor enclosed in a sealed cavity that can be maintained at constant temperature and pressure. Usually, the hygrometer measures the vapor pressure above solution, soil, or plant-tissue samples by the manipulation of energy flow to and from a thermocouple (Savage, 1982; Savage and Cass, 1984).

7. THEORY OF THE PRESSURE-CHAMBER APPARATUS

When a transpiring leaf is severed (at the petiole), the xylem sap recedes. Pressurizing the leaf until the water just returns to the cut surface gives a measure of the hydrostatic pressure in the xylem (Scholander et al., 1965). The pressure-chamber apparatus is in fact analogous to the pressure-membrane (sometimes referred to as pressure-plate) apparatus used mainly in soil physics (Passioura, 1980) to measure matric potentials of soils and other materials. The two methods are analogous in that soil matric potential is measured, but in the case of the Scholander pressure chamber apparatus, the matric potential in the apoplast (or cell wall) is measured. Provided that the osmotic or solute potential (Ψ_s) of the apoplastic water is close to 0 MPa, the equation:

$$\Psi = \Psi_p + \Psi_s + \Psi_m \quad \text{reduces to} \quad \Psi = \Psi_p + \Psi_m,$$

where Ψ_p is the pressure applied to balance Ψ_m, the matric potential of the apoplast, resulting in a total potential Ψ of 0 MPa. Hence, the matric potential of the apoplast, Ψ_m, is equal to $-\Psi_p$. It is usually assumed that the matric potential of the apoplast is equal to the total water potential of the symplast, so that the pressure chamber then measures the total water potential of the leaf.

ACKNOWLEDGMENTS

This work was sponsored by the Foundation for Research Development, the Department of Agriculture, and the University of Natal South Africa.

REFERENCES

Babcock, K.L. 1963. Theory of the chemical properties of soil equilibrium. Hilgardia 34:417-542.

Bolt, G.H. and M.J. Frissel. 1960. Thermodynamics of soil water. Netherlands Journal of Agricultural Science 8:57-78.

Bolt, G.H., S. Iwata, A.J. Peck, P.A.C. Raats, A.A. Rode, G. Vachaud, and A.D. Voronin. 1975. Soil physics terminology. Bulletin of the International Soil Science Society 48:26-36.

Dainty, J. 1976. Water relations of plant cells. In Encyclopedia of Plant Physiology, volume 2: Transport in Plants II: Part A Cells, p 12-35.

Passioura, J.B. 1980. The meaning of matric potential. Journal of Experimental Botany 31:1161-1169.

Rose, D.A. 1979. Soil water: quantities, units, and symbols. Journal of Soil Science 30:1-15.

Salisbury, F.B. and C.W. Ross. 1991. Plant Physiology, Fourth Edition. Wadsworth Publishing Company, Belmont, California.

Savage, M.J. 1978. Water potential terms and units. Agrochemophysica 10:5-6.

Savage, M.J. 1979. Use of the international system of units in the plant sciences. HortScience 15:492-495

Savage, M.J. 1982. Measurement of water potential using thermocouple hygrometers. Unpublished Ph.D. thesis, University of Natal, Pietermaritzburg, South Africa. 162 p.

Savage, M.J. and A. Cass. 1984. Measurement of water potential using *in situ* thermocouple hygrometers. Advances in Agronomy 37: 73-126.

Scholander, P.F., H.T. Hammel, E.D. Bradstreet, and E.A. Hammingsen. 1965. Sap pressure in vascular plants. Science 148:339-346.

Slatyer, R.O. 1967. Plant-Water Relationships. Academic Press, New York.

Spanner, D.C. 1973. The components of the water potential in plants and soils. Journal of Experimental Botany 24:816-819.

Taylor, S.A. and G.L. Ashcroft. 1972. Physical Edaphology. W.H. Freeman and Company, San Francisco.

CONSULTANTS

Keith L. Bristow
CSIRO
Townsville, Queensland, Australia

Gaylon S. Campbell
Washington State University
Pullman, Washington

Alfred Cass
CSIRO
Glen Osmond, South Australia, Australia

Jack Dainty
University of Toronto
Toronto, Ontario, Canada

George C. Green
Water Research Commission
Pretoria, South Africa

Frank B. Salisbury
Utah State University
Logan, Utah

5

SOLUTIONS (IONIC RELATIONS)

Jack Dainty[1]
Department of Botany
University of Toronto
Toronto, Ontario M5S 1A1
Canada

Many events in plants involve movements of substances as gases or as solutes, molecular or ionic, dissolved in liquids, typically water. The following are recommended symbols and units to be used in discussion of these movements.

1. ABBREVIATIONS USED AS SUBSCRIPTS AND SUPERSCRIPTS

j	subscript for any component in a mixture
in, out i, o co, oc cv, vc	subscripts or superscripts denoting the direction of a process; e.g., a flux from cytoplasm (c) to vacuole (v) or from cytoplasm (c) to external medium (o).
w	subscript for water
P	subscript for pressure
V	subscript for volume
s	subscript for solute
π	subscript for osmotic potential
τ or m	subscript for matric potential
cw	subscript for cell wall
o, i	superscript used for outside or inside of compartment (e.g., a cell).

A bar over a symbol usually means **average**; e.g., $\bar{C}$ = average concentration, but it can also mean **partial molar** as in $\bar{V}_j$.

[1]Current address is: Jack Dainty, Mas Tourriere, F-34270, Cazevieille, France.

2. THE TABLES

Table 1. Recommended Units for Concentrations [a, b] *(Symbol C)*

solids in solids	$mol \cdot kg^{-1}$ or mol mol^{-1} or kg kg^{-1}
solids in liquids	$mol \cdot m^{-3}$ or kg m^{-3} (SI units) $mol \cdot L^{-1}$ = M = molar concentration (not recommended) $mol \cdot kg^{-3}$ = m = molal concentration (not recommended). $kg \cdot L^{-1}$ (acceptable when not a pure substance; avoid $mg \cdot mL^{-1}$, etc.)
solids in gases [c]	$mol \cdot m^{-3}$ or $mol \cdot mol^{-1}$ or $kg \cdot m^{-3}$
liquids in liquids	$mol \cdot m^{-3}$ or $mol \cdot mol^{-1}$ or $kg \cdot m^{-3}$ (SI units) $mol \cdot L^{-1}$ or $L \cdot L^{-1}$ or $kg \cdot L^{-1}$ (acceptable with SI units)
liquids in gases [c]	$mol \cdot m^{-3}$ or $kg \cdot m^{-3}$ or $mol \cdot mol^{-1}$
gases in gases [c]	$mol \cdot mol^{-1}$ or $mol \cdot m^{-3}$ or $m^3 \cdot m^{-3}$ Avoid parts per million, parts per billion, etc.; $L \cdot L^{-1}$ (e.g., $\mu L \cdot L^{-1}$) is acceptable.
gases in liquids [c]	Same as gases in gases.
gases in solids [c]	$mol \cdot mol^{-1}$ or $mol \cdot kg^{-1}$ or $m^3 \cdot kg^{-1}$ ($L \cdot kg^{-1}$ is acceptable)

[a] This table was prepared by F.B. Salisbury in response to a suggestion of T.W. Tibbitts.
[b] Use moles for pure substances; otherwise, use kilograms.
[c] When volume (m^3 or L) is used for gases, temperature and pressure must be specified.

Table 2. Recommended Symbols and Units for Plant Ionic (Solution) Relations

Parameter	*Symbol*	Unit
concentration	C_j	$mol \cdot m^{-3}$
	M_j	$mol \cdot L^{-1}$ (M = molarity; discouraged)
	m_j	$mol \cdot kg^{-1}$
mole fraction	x_j (or X_j)	$mol \cdot mol^{-1}$ (dimensionless)
activity	a_j	Same units as corresponding concentration
activity coefficient	f_j	dimensionless; used when concentration is expressed in $mol \cdot m^{-3}$, $mol \cdot L^{-1}$, or $mol \cdot kg^{-1}$; defined by $a_j = f_j C_j$ etc.
	v_j	dimensionless; defined by: $a_j = v_j X_j$

Continued

Table 2. Recommended Symbols and Units for Plant Ionic Relations (continued)

Parameter	*Symbol*	Unit
partial molar volume	$\bar{V}_j = \left(\frac{\partial V}{\partial n_j}\right)$	$m^3 \cdot mol^{-1}$
amount of pure substance	n_j	mol
chemical potential	μ_j	$J \cdot mol^{-1}$
electrochemical potential	$\tilde{\mu}_j$	$J \cdot mol^{-1}$

Note: Chemical potential for nonelectrolyes or water is given by:

$$\mu_j = \mu_j^* + RT \ln a_j + P\bar{V}_j$$

for component j. Electrochemical potential has come to mean the chemical potential of an ion and is expressed, for ion j, by:

$$\tilde{\mu}_j = \tilde{\mu}_j^* + RT \ln a_j + P\bar{V}_j + z_j F \psi.$$

In these formulae for chemical potential and electrochemical potential, R is the gas constant (8.314 J mol^{-1} K^{-1}), T is the absolute (kelvin) temperature, and F is the Faraday constant (9.648 x 10^4 $C \cdot mol^{-1}$). P is the pressure in Pa, $\bar{V}_j$ the partial molar volume of j in $m^3 \cdot mol^{-1}$, z_j is the algebraic valency (see entry below), ψ is the electrical potential in volts (v), and a_j is the activity in appropriate units. The chemical and electrochemical potentials in the standard states are given by μ_j^* and $\tilde{\mu}_j^*$. For nonelectrolyte and ionic solute species, the pressure term, $P\bar{V}_j$ is usually negligibly small.

Parameter	*Symbol*	Unit
electrical potential	ψ *(or E)*	V
electrical potential difference	$\Delta\psi$ *(or* ΔE*)*	V

Note: Electrical potential difference is often symbolized as V or E; for example, the membrane potential $\psi^i - \psi^o = \Delta\psi$ is often written as V_m or E_m or V^{io} or E^{io}. Strictly, however, the symbol E should be reserved for electromotive force; for example, the Nernst potential E_j for an ion (see next entry). E_j can be considered an electromotive force.

Parameter	*Symbol*	Unit
Nernst potential	E_j	V (volt)

$$E_j = \left(\frac{RT}{z_j F}\right) \ln\left(\frac{a_j^o}{a_j^i}\right)$$

Note: The symbols are explained for the chemical or electrochemical potential. The superscripts o and i refer to the outside and inside phases.

Parameter	*Symbol*	Unit
algebraic valency	z_j	dimensionless

Note: Here used in the sense of the number of electron charges per ion. The symbols z_j^+ and z_j^- are often used for the charges carried by a cation and an anion, respectively.

Continued

Table 2. Recommended Symbols and Units for Plant Ionic Relations (continued)

Parameter	*Symbol*	Unit
mobility	u_j	either: $m^2 \cdot s^{-1} \cdot V^{-1}$ or: $m \cdot mol \cdot s^{-1} \cdot N^{-1}$
Note: The units will depend on whether the driving force is considered as a voltage gradient: $m \cdot s^{-1}/(V \cdot m^{-1})$, or as a force per unit amount of ions: $m \cdot s^{-1}/(N \cdot mol^{-1})$.		
electric current	I	A (amp)
electrical capacitance	C	F (farad)
charge	Q	C (coulomb)
electrical resistance	R	Ω (ohm)
electrical conductance	G	S (siemen)
specific electrical conductance	g, g_j	$S \cdot m^{-2}$
Note: The symbols g, g_j can be considered either as a slope conductance, $\partial I / \partial E$, or as a chord conductance; e.g., $I_j / (E_m - E_j)$.		
transport number	t_j	dimensionless
Note: The transport number, t_j, is the fraction of an electric current in a solution, or passing through a membrane, for instance, carried by ion j: $\Sigma t_j = 1$.		
flux	ϕ, J	$mol \cdot m^{-2} \cdot s^{-1}$
influx	ϕ_{in}, ϕ_i, J_{in}	
efflux	ϕ_{out}, ϕ_o, J_{out}	
other (specific) fluxes	ϕ_{oc}, ϕ_{co}, ϕ_{cv}, ϕ_{vc}, etc.	
Note that the term *flux* is used in transport studies for amount crossing unit area per second. Thus terms such as "rate of flux" or "flux density" are incorrect and should not be used.		
permeability coefficient	P	$m \cdot s^{-1}$
diffusion coefficient	D	$m^2 \cdot s^{-1}$
partition coefficient	K	dimensionless
Michaelis constant	K_m	$mol \cdot m^{-3}$ (or $mol \cdot L^{-1}$, molarity, M, but should be avoided)
rate constant	k	s^{-1} $m^3 \cdot mol^{-1} \cdot s^{-1}$, etc.
velocity (e.g., of ions)	v	$m \cdot s^{-1}$
velocity (maximum rate of transport)	V_{max}	$mol \cdot m^{-2} \cdot s^{-1}$
velocity (of reaction)	v	$mol \cdot s^{-1}$

Continued

Table 2. Recommended Symbols and Units for Plant Ionic Relations (continued)

Parameter	*Symbol*	Unit
velocity (maximum of enzymatically controlled reaction)	V_{max}	$mol \cdot s^{-1}$
generalized force	X	usually $J \cdot m^{-1} \cdot mol^{-1}$ (or $N \cdot mol^{-1}$)
generalized conductance coefficient	L	$mol \cdot m^{-2} \cdot s^{-1}$/"Force"
quantity of substance	Q_j	mol
quantity of isotope	Q_j^*	appropriate units; e.g., the becquerel, Bq (becquerels are expressed as s^{-1})
specific activity	S_j	appropriate units e.g., $Bq \cdot mol^{-1}$

CONSULTANTS: see next chapter.

6

WATER RELATIONS

Jack Dainty
Department of Botany
University of Toronto
Toronto, Ontario M5S 1A1
Canada

In organisms in general, but particularly in plants, the movement of water is of special importance. Its movement by diffusion or bulk flow follows the same thermodynamic principles as the movement of other substances, but its prevalence in living systems has provided an impetus for special study and for a special set of terms and measurements. At present, authors of technical papers use one or the other of the two ways of expressing the solute effect on the chemical potential of water: the *osmotic pressure* or *osmotic (solute) potential.* **Osmotic pressure**, which expresses the effect of solutes as a positive number with dimensions of pressure, is the traditional approach, but **osmotic potential**, which considers the solute effect as a component of the water potential, expresses the effect as a negative number, usually also with dimensions of pressure (see Chapter 4 on thermodynamics). In spite of tradition, the concept of *potentials* seems most logical from the standpoint of thermodynamics, which is the basis for the universally accepted concept of water potential. In the following summary of recommended symbols and units, both approaches are presented, but the use of *potentials* is strongly encouraged.

1. THE TABLES

The basic recommended terms, symbols, and units are summarized in Table 1. The discussion following the table expands the basic potential terms and gives the traditional terms as well. Table 2 then summarizes the terms, symbols, and units considered in the discussion. Table 3 presents other terms used in discussions of plant water relations.

Table 1. Recommended Terms, Symbols, & Units for Plant Water Relations

Parameter	*Symbol*	Unit
chemical potential of water	μ_w	$J \cdot mol^{-1}$
water potential	Ψ	Pa
components of water potential: pressure potential (= hydrostatic pressure), solute potential, and matric potential.	Ψ_p (P), $\Psi_s(\Psi_\pi)$, & Ψ_m	Pa (Pascal) (0.1 MPa = 1 bar; MPa is usually most appropriate, but kPA may also be used; $J \cdot mol^{-1}$ or $J \cdot kg^{-1}$ can also be used.)

Water potential is defined by $\Psi = \dfrac{(\mu_w - \mu_w^*)}{\bar{V}_w}$ where μ_w^* = the chemical potential (units $J \cdot mol^{-1}$) of pure water at atmospheric pressure and at the same temperature as the system under consideration, and $\bar{V}_w$ = the partial molar volume of water equals (18 000 $mm^3 \cdot mol^{-1}$).

As discussed in Chapter 4 on thermodynamics, division of the reference chemical potential of water (μ_w - μ_w*), which has units of energy (J mol^{-1}), by the partial molar volume of water produces units equivalent to those of pressure. This is illustrated by the following conversions (see Table 2 in Chapter I of this book):

$$J \cdot mol^{-1} = N \cdot m \cdot mol^{-1} = m^2 \cdot kg \cdot s^{-2} \cdot mol^{-1}$$

$$\frac{m^2 \cdot kg \cdot s^{-2} \cdot mol^{-1}}{m^3 \cdot mol^{-1}} = kg \cdot s^{-2} \cdot m^{-1} = N \cdot m^{-2} = Pa$$

Water potential is often written as a sum of partial potentials (see discussion in Chapter 4):

$$\Psi = \Psi_p + \Psi_s \; (+ \Psi_m)$$

where

Ψ_P = the **pressure potential** (actual hydrostatic **pressure**, P; positive or negative),

Ψ_s = the **osmotic potential** or **solute potential** (often written Ψ_π; sometimes, incorrectly, as π; *always negative*), and

Ψ_m = the **matric potential** (considered by some not to be a valid component of Ψ; *always negative*).

The other usual way of writing the equation for water potential is:

$\Psi = P - \pi \; (- \tau)$

where

Ψ and P = water potential and pressure as above,

π = **osmotic pressure** (Ψ_s = $-\pi$; unit Pa, sometimes $osmol \cdot kg^{-1}$), and

τ = **matric potential** (Ψ_m = $-\tau$; unit Pa).

Note that **osmotic pressure** is numerically equivalent to **solute potential** but expressed as a positive instead of a negative value; there are also positive and negative versions of **matric potential** ($\Psi_m = -\tau$). Use of osmotic pressure instead of osmotic potential is traditional and optional, but such a usage does not emphasize that the parameter is one component of water potential. It is recommended that osmotic potential be used in plant physiology.

Osmotic pressure is, occasionally in the plant literature and often in the animal literature, expressed in osmol kg^{-1}. A value of X osmol kg^{-1} means RTX Pa (in pressure units); that is, the aqueous solution under consideration has the same osmotic pressure as an *ideal* solution of molality X mol kg^{-1}.

The **matric potential** is the (negative) effect of the solid (and gaseous) phases on the water potential. Its reality is often doubted, particularly if P and π are correctly interpreted. (See J.B. Passioura. 1980. The meaning of matric potential. J. of Exper. Botany 31:1161-1169; F.B. Salisbury and C.W. Ross. 1991. Plant Physiology, Fourth Edition, Wadsworth Pub. Co., Inc; Belmont, California.)

Table 2. Summary of Terms Defined and Discussed in Table 1.

Parameter	*Symbol*	Unit
water potential	Ψ (usually negative, can be positive)	Pa $J \cdot kg^{-1}$
pressure potential (or **pressure**)	Ψ_p or P (can be + or -)	Pa $J \cdot kg^{-1}$
solute or **osmotic potential**	Ψ_s (Ψ_π; always negative)	Pa $J \cdot kg^{-1}$
osmotic pressure	π ($\Psi_s = -\pi$; π is always positive)	Pa $J \cdot kg^{-1}$
matric potential	Ψ_m (Ψ_τ; $\Psi_m = -\tau$; Ψ_m is always negative; τ is always positive)	Pa $J \cdot kg^{-1}$

Table 3. Other Terms Used in Discussions of Plant Water Relations.

Parameter	*Symbol*	Unit
volume flux	J_v	$m \cdot s^{-1}$
solute flux	ϕ_s, J_s	$mol \cdot m^{-2} \cdot s^{-1}$
solute permeability	P_s or ω_s	$m \cdot s^{-1}$ $mol \cdot m^{-2} \cdot s^{-1} \cdot Pa^{-1}$

Continued

Table 3. Other Terms Used in Discussions of Plant Water Relations (continued)

Parameter	*Symbol*	Unit
	Note: ω_s is defined by the equation: flux = $\omega_s RT\Delta C_s$ and is thus given by $\omega_s RT = P_s$.	
hydraulic conductance	L_p	$m \cdot s^{-1} \cdot Pa^{-1}$
	Note: This refers usually to the cell membrane and is often incorrectly called hydraulic conductivity. It is not normalized to unit thickness of the barrier. The term hydraulic conductivity is correctly used when, for example, referring to the conductivity to water of the cell wall material; the same symbol is usually used, but the units are $m^2 \cdot s^{-1} \cdot Pa^{-1}$.	
hydraulic resistance	R (= $1/L_p \cdot A$, where A = area)	$Pa \cdot s \cdot m^{-3}$
	Note: Refers to cells, tissues, organs, or entire plants.	
diffusivity	D	$m^2 \cdot s^{-1}$
	Note: Denotes the "speed" at which changes in water potential propagate within tissues and incorporates cell and wall conductances to water and their storage capacities.	
diffusional permeability for water	P_d	$m \cdot s^{-1}$
reflection coefficient	σ	dimensionless
	Note: For a membrane that is leaky toward the solute (i.e., not semipermeable), volume flux, J_v, is given, not by $J_v = L_p(\Delta P - \Delta\pi)$, but by $J_v = L_p(\Delta P - \sigma\Delta\pi)$. The reflection coefficient, σ, expresses the ratio between the apparent osmotic pressure and the thermodynamic osmotic pressure; σ, is always less than one, and usually $0 \leq \sigma \leq 1$.	
volume modulus of elasticity (for a cell)	ϵ ϵ is defined as $V(dP/dV)$	Pa
non-osmotic volume	b	m^3 (often given as % or $m^3 \cdot m^{-3}$)
osmotic coefficient	ϕ	dimensionless
viscosity	η	$N \cdot s \cdot m^{-2}$
extensibility (of cell wall)	m	$m^2 \cdot N^{-1} \cdot s^{-1}$
kinematic viscosity	η_T	$m \cdot s^{-1}$
thickness of unstirred layer	δ	m
surface tension	T	$N \cdot m^{-1}$

CONSULTANTS FOR THE CHAPTERS BY JACK DAINTY

Mary A. Bisson
SUNY - Buffalo
Buffalo, New York

Julian Collins
University of Liverpool
Liverpool, England

John Cram
University of Newcastle
Newcastle, England

Robert F. Davis (deceased)
Rutgers University
Newark, New Jersey

Dieter Jeschke
Estenfeld, Germany

Betty L. Klepper
USDA-ARS
Pendleton, Oregon

William J. Lucas
University of California
Davis, California

Enid A. C. MacRobbie
Botany School
Cambridge, England

E. Marrè
Universita' degli di Milano Studii
Milano, Italy

John A. Milburn
The University of New England
Armidale, Australia

Michael G. Pitman
CSIRO
Dickson, Australia

Ronald J. Poole
McGill University
Montreal, Québec, Canada

Leonora Reinhold
Hebrew University of Jerusalem
Jerusalem, Israel

Roger M. Spanswick
Cornell University
Ithaca, New York

Ernst Steudle
Universität Bayreuth
Bayreuth, Germany

Michel Thellier
Faculté des Sciences de Rouen
Mont-Saint-Aignan, France

Alan Walker
University of Sydney
Sydney, Australia

7

ENERGY TRANSFER

Frank B. Salisbury
Plants, Soils, and Biometeorology Department
Utah State University
Logan, Utah 84322-4820 U.S.A.

Michael J. Savage
Department of Agronomy
University of Natal
Pietermaritzburg 3201
Republic of South Africa

An excellent example of plant biophysics is the application of physical principles to understand the energy exchange between a plant and its environment. Here, we present a summary table of terms, symbols, and units that are appropriate for this endeavor followed by some equations that are often used.

1. TERMS, SYMBOLS, AND UNITS APPROPRIATE IN ENERGY-TRANSFER STUDIES

The symbols in Table 1 are mostly arranged alphabetically, with Roman letters first, then Greek letters (but some parameters have alternative Roman or Greek symbols).

Table 1. Terms, Symbols, and Units.

Parameter	***Symbol***	**Unit**
absorption coefficient, like reflection coefficient, need not have λ (wavelength) specified	a	unitless
absorptance or absorptivity	a	unitless ($J \cdot J^{-1}$)
heat energy storage	B	$J \cdot kg^{-1}$, $W \cdot m^{-2}$
boundary layer, superscript	bl	
conduction, superscript	c	

Continued

Table 1. Terms, Symbols, and Units (continued)

Parameter	*Symbol*	Unit
volumetric heat capacity (at constant volume)	c_v	$J \cdot m^{-3} \cdot {}^\circ C^{-1}$ or $J \cdot m^{-3} \cdot K^{-1}$
specific heat capacity of dry air (at constant pressure)	c_p	$J \cdot kg^{-1} \cdot {}^\circ C^{-1}$ or $J \cdot kg^{-1} \cdot K^{-1}$
diffusion coefficient of species *j*	D_j	$m^2 \cdot s^{-1}$
emissivity or emittance in infrared region, for example	ϵ_{IR} and E_{IR}	unitless
water evaporation site, superscript	e	
radiant energy, kinetic energy	E	J
leaf conductance	g_{leaf}	$m \cdot s^{-1}$
coefficient, convective transfer coefficient, heat energy transfer coefficient heat energy convection	h_c	$W \cdot m^{-2} \cdot {}^\circ C^{-1}$ or $W \cdot m^{-2} \cdot K^{-1}$
Planck's constant	h	$=6.626 \times 10^{-34}$ $J \cdot s \cdot photon^{-1}$
a quantum of radiant energy	$h\nu$	J
sensible heat energy transfer	H	$W \cdot m^{-2}$
heat energy, subscript	h	
infrared	*IR* (near infrared: 800 to 3000 nm) (far infrared: 3000 to 70 000 nm)	
foliar absorption coefficient	k	unitless
thermal conductivity coefficient of region *j*	K^j	$W \cdot m^{-2} \cdot {}^\circ C^{-1}$ or $W \cdot m^{-2} \cdot K^{-1}$
eddy diffusion coefficient of gaseous species	K_j	$m^2 \cdot s^{-1}$
metabolic heat energy	M	$W \cdot m^{-2}$, $J \cdot kg^{-1}$
photosynthetic irradiance (photosynthetically active radiation)	*PAR*	$W \cdot m^{-2}$
photosynthetic photon flux (photosynthetically active radiation, photon basis)	*PPF*	$mol \cdot m^{-2} \cdot s^{-1}$ (moles of photons per square meter second) (usually $\mu mol \cdot m^{-2} \cdot s^{-1}$) (also: $mol \cdot m^{-2} \cdot d^{-1}$)
partial pressure of gaseous species *j*	P_j	Pa (usually kPa)

Continued

Table 1. Terms, Symbols, and Units (continued)

Parameter	*Symbol*	Unit
vapor pressure, leaf and air	e_l, e_a	Pa
net irradiance	Q	$W \cdot m^{-2}$
Universal gas constant	R	$8.3143\ J \cdot mol^{-1} \cdot K^{-1}$ $8.3143\ m^3 \cdot Pa \cdot mol^{-1} \cdot K^{-1}$ $8.3143 \times 10^{-6}\ m^3 \cdot MPa \cdot mol^{-1} \cdot K^{-1}$ $(1.987\ cal \cdot mol^{-1} \cdot K^{-1})$ $(0.083143\ L \cdot bar \cdot mol^{-1} \cdot K^{-1})$
reflection coefficient	r	unitless
reflectivity (λ indicates wavelength reflected)	r_λ or $r(\lambda)$	unitless
resistance for gaseous diffusion for species *j*	r_j	$s \cdot m^{-1}$
boundary-layer resistance (a for aero-dynamic)	r_a	$s \cdot m^{-1}$
diffusive resistance within a leaf	r_l or r_ℓ	$s \cdot m^{-1}$
relative humidity	*RH*	percent (%)
solar irradiance; i.e., global irradiance	S	$W \cdot m^{-2}$
turbulent air, superscript	*ta*	
transpiration, superscript	T	
temperature	T	K (°C)
kinetic energy per amount of substance	U	$J \cdot mol^{-1}$
ultraviolet	*UV*(*UV-A* 320 to 400 nm) (*UV-B* 280 to 320 nm) (*UV-C* ⟨ 280 nm)	
specific latent heat of vaporization; tranpiration or condensation; specific latent heat of fusion	V or L	$J \cdot kg^{-1}$, $W \cdot m^{-2}$
velocity, wind speed	v	$m \cdot s^{-1}$
water, water vapor, subscripts	*w*, *wv*	
distance	x, or δ (delta)	m

Continued

Table 1. Terms, Symbols, and Units (continued)

Parameter	*Symbol*	Unit
psychrometric constant	γ (gamma)	$Pa \cdot K^{-1}$ Typical value is 66.6 $Pa \cdot K^{-1}$ (at 20°C and 100 kPa: sea level)
thickness of air boundary layer	δ^{bl} (delta)	m (usually mm)
difference or change in the quantity that follows	Δ (delta)	
emissivity or emittance in infrared region (as example)	ϵ_{IR} (epsilon)	unitless ($J \cdot J^{-1}$)
wavelength of radiation	λ (lambda)	nm
wavelength corresponding to the maximum absorption coefficient in an absorption band or to the maximum photon (or energy) emission in an emission spectrum	λ_{max} (lambda)	nm
frequency of electromagnetic radiation	ν (nu)	s^{-1}, Hz (hertz)
density of dry (unsaturated) air	ρ (rho)	$kg \cdot m^{-3}$
Stefan-Boltzmann constant	σ (sigma) or δ (delta)	$=5.673 \times 10^{-8}\ W \cdot m^{-2} \cdot K^{-4}$

2. SOME EQUATIONS USED IN HEAT-TRANSFER STUDIES

A. The Energy Balance Equation for a Leaf Surface (all values can be expressed as watts per square meter: $W \cdot m^{-2}$):

$$Q + H + V + B + M + A = 0$$

where

Q = net irradiance (positive if leaf is radiating less energy than the radiant energy absorbed from its surroundings),

H = sensible heat flux transfer (includes conduction and convection; negative if leaf loses more heat energy than it gains),

V = latent heat flux; the transpiration term (negative when water is vaporizing; positive when condensing or freezing),

B = storage flux (positive when leaf temperature is increasing),

M = metabolism and other factors (positive when heat is produced), and

A = advected heat flux from leaf to air (positive for advection from air to leaf; advection is the horizontal flow of air—i.e., wind).

At constant leaf temperature and ignoring metabolism and advection (which could be important but is difficult to measure): $Q + H + V = 0$.

B. Radiant Energy Flux Absorbed by a Leaf Surface (Q_{abs}; $W \cdot m^{-2}$):

$Q_{abs} = \epsilon Q_{PAR} + \epsilon' Q_{th}$

where

ϵQ_{PAR} = total absorbed irradiance in the PAR region ($W \cdot m^{-2}$),

$\epsilon' Q_{th}$ = total absorbed (thermal) irradiance outside PAR region ($W \cdot m^{-2}$),

and

ϵ and ϵ' = leaf emissivities in the two spectral regions.

C. Radiant Energy Flux from a Leaf (or any) Surface ($Q\epsilon$; $W \cdot m^{-2}$):

$Q\epsilon = \epsilon \sigma T^4$

where

$Q\epsilon$ = Radiant energy flux ($W \cdot m^{-2}$),

ϵ = emissivity (about 0.98 for leaves at growing temperatures),

σ = Stefan-Boltzmann constant (5.673×10^{-8} $W \cdot m^{-2} \cdot K^{-4}$), and

T = absolute temperature of the leaf (K)

This **Stefan-Boltzmann Law** is applied in the next equation.

D. Net Irradiance at a Leaf Surface (Q; $W \cdot m^{-2}$):

Energy flux emitted by a leaf (Stefan-Boltzmann law) is subtracted from the absorbed radiant energy flux (Q_{abs}):

$Q = Q_{abs} - \epsilon_{IR} \sigma T^4$

where

Q = energy flux ($W \cdot m^{-2}$)

Q_{abs} = absorbed energy flux ($W \cdot m^{-2}$), and

ϵ_{IR} = emissivity or absorptivity of the leaf for long-wave (thermal) radiation; typically about 0.95 for living leaves at normal temperatures (same as ϵ' above).

Often, the above equation is written (see Monteith and Unsworth, 1990):

$Q = I_s - rI_s + L_{env} - \epsilon_{IR} \sigma T^4$

where

I_s = the solar irradiance incident at the leaf surface ($W \cdot m^{-2}$),

r = the leaf surface reflection coefficient (decimal fraction), and

L_{env} = the environmental longwave irradiance incident at the leaf surface ($W \cdot m^{-2}$).

E. Sensible Energy Flux Transfer by Convection at a Leaf Surface (H; $W \cdot m^{-2}$):

$$H = \frac{c_p \rho (T_a - T_l)}{r_a} = \frac{c_p \cdot \rho \cdot \Delta T}{r_a} = c_p \cdot \rho \cdot \Delta T \cdot g_a$$

where

T_a = air temperature (K or °C),

T_l = leaf temperature (K or °C),

ΔT = $T_a - T_l$,

c_p = specific heat capacity of dry (unsaturated) air (≈ 1000 $J \cdot kg^{-1} \cdot K^{-1}$) at constant pressure,

ρ = density of dry air (1.205 $kg \cdot m^{-3}$ at 20 °C and 100 kPa),

r_a = boundary-layer resistance ($s \cdot m^{-1}$), and

g_a = boundary-layer conductance ($m \cdot s^{-1}$).

The **convective transfer coefficient** (***h***$_c$; $W \cdot m^{-2} \cdot K^{-1}$), also called the **heat transfer coefficient** (proportional to the reciprocal of the boundary layer resistance), may be used to calculate sensible energy transfer ***H*** ($W \cdot m^{-2}$):

$$h_c = \frac{c_p \rho}{r_a}$$

$$H = \frac{c_p \rho \cdot \Delta T}{r_a} = c_p \rho \cdot g_a \cdot \Delta T$$

$$H = h_c \Delta T.$$

F. Latent Energy Flux of Water Vapor at a Leaf Surface (*V*; $W \cdot m^{-2}$), the Transpiration Term:

$$V = \frac{(e_\ell - e_a) c_p \rho}{\gamma (r_l + r_a)} = \frac{\Delta e \cdot c_p \rho}{\gamma (r_l + r_a)} = \frac{\Delta e \cdot c_p \rho}{\gamma (\frac{1}{g_l} + \frac{1}{g_a})}$$

where

e_l = vapor pressure in the leaf; i.e., within the substomatal cavity (Pa),

e_a = vapor pressure of the air (Pa),

r_a = boundary layer resistance (in air) ($s \cdot m^{-1}$),

r_l = diffusive resistance within the leaf ($s \cdot m^{-1}$),

γ = psychrometric constant (typically 66.6 $Pa \cdot K^{-1}$), and

g_l and g_a = leaf and boundary-layer conductivities ($m \cdot s^{-1}$), respectively.

REFERENCES

Campbell, Gaylon S. 1977. An Introduction to Environmental Biophysics. Springer-Verlag, New York, Heidelberg, Berlin. 159 p.

Gates, David M. and LaVerne E. Papian. 1971. Atlas of Energy Budgets of Plant Leaves. Academic Press, London and New York. 279 p.

Gates, David M. 1968. Transpiration and leaf temperature. Annual Review of Plant Physiology 19:211-238.

Larcher, Walter. 1995. Physiological Plant Ecology, Third Edition. Springer-Verlag, Berlin, Heidelberg, New York. (Translated by Joy Wieser) 506 p.

Monteith, J.L. and M.H. Unsworth. 1990. Principles of Environmental Physics. Edward Arnold: London, 291 p.

Nobel, Park S. 1983. Biophysical Plant Physiology and Ecology. W.H. Freeman and Company, San Francisco. 608 p. [The symbols and units used in this chapter were modified from those in this text book.]

Raschke, Klaus. 1960. Heat transfer between the plant and the environment. Annual Review of Plant Physiology 11:111-126.

CONSULTANTS

Donald T. Krizek
USDA Agricultural Research Service
Beltsville, Maryland

John C. Sager
John F. Kennedy Space Center
Kennedy Space Center, Florida

8

PHLOEM TRANSPORT

Donald R. Geiger
Department of Biology
University of Dayton
Dayton, Ohio 45469-2320 U.S.A.

Aart J.E. van Bel
Botanisches Institut 1
Justus-Liebig Universität
Senckenbergstrasse 17
D-35390 Giessen, Germany

In this chapter terms are defined and SI units are presented when appropriate.

Table 1. Some Terms and Units Used in the Study of Phloem Transport.

Term	Description of Concept	Units
BASIC AND DESCRIPTIVE TERMS:		
photoassimilates	Organic compounds produced by photosynthetic carbon fixation.	
translocation	Long distance transport of solutes through sieve tubes or other structures specialized for longitudinal transport.	
allocation [a] **(partitioning)**	Flow of photoassimilates into various compartments or biochemical pathways within source and sink regions. In source organs, carbon is allocated to various uses including export. In a sink organ, carbon enters into compartments or is used for synthesis, storage, or energy metabolism.	
partitioning [a] **(allocation)**	Distribution of translocated photoassimilates among sinks.	
pressure flow hypothesis	The theory of osmotically-driven pressure flow within the phloem; pressure builds up in the sieve element-companion cell complexes in the source as water moves into these sieve element members in response to high solute concentrations therein; pressures within the sieve element members are less in the sink regions as solutes exit from the phloem. First proposed by E. Münch in 1930.	

Continued

Table 1. Some Terms and Units Used in the Study of Phloem Transport (continued)

Term	Description of Concept	Units
pressure flow	As applied to phloem transport, the mass flow of water and solutes along a pressure gradient.	
osmotically generated flow	Flow that arises from negative osmotic potential within sieve tubes (phloem translocation) or xylem vessels (root pressure, root exudation, guttation).	
mass flow	Flow of solute along with solvent.	
sieve-element/ companion-cell complex	Cellular complex interconnected by specialized unilaterally branched plasmodesmata; the complex acts as an integrated physiological unit in photoassimilate transport.	
apoplast	The interconnecting cell walls and water-filled xylem elements in a plant through which water and dissolved solutes can move freely. In a sense, the "dead" part of a plant but, from a functional standpoint, excluding the suberin-filled Casparian strips.	
symplast	The interconnected, through plasmodesmata, protoplasts of a plant. Some authors would exclude the central vacuoles. In a sense, the "living" part of a plant.	
QUANTITATIVE OR MEASUREMENT TERMS		
translocation profile	A plot of solute concentration versus distance (spatial profile) or solute concentration at a particular location versus time (**temporal profile**).	
bidirectional transport	Simultaneous transport of solutes in opposite directions in the same file of sieve elements. The process has not been demonstrated in the sense of the definition. Bidirectional transport may occur under some circumstances, for example, in a file of cells as a result of cytoplasmic streaming.	
source	A region in which net flux of solute into elements is sufficient to cause net export from them.	
sink	A region in which there is net efflux of solutes and water from sieve elements, resulting in import into them.	
phloem loading	The process by which products of carbon assimilation enter into the sieve-element/companion-cell complex. The process may be: a) apoplastic as when solutes are transported across the plasma membrane of the sieve-element/companion-cell complex, b) symplastic as when solutes are transported from surrounding mesophyll into the sieve element companion cell complex through symplastic connections, c) symplastic and apoplastic when both pathways operate in parallel or alternatively.	
phloem unloading	The processes that bring about the net transport of water and solutes from sieve elements in sink organs into the surrounding sink tissues. Passage out of the sieve elements may be symplastic or apoplastic or both, depending on the nature of the sink.	

Continued

Table 1. Some Terms and Units Used in the Study of Phloem Transport (continued)

Term	Description of Concept	Units
FLOW TERMS:		
phloem export rate	Rate of phloem translocation of a specified solute out of a source organ. A basis for comparison such as per leaf, per area, or per plant should be specified.	$mg \cdot s^{-1}$ or $mol \cdot s^{-1}$
phloem import rate	Rate of entry of a specified solute through phloem into a sink organ. The rate should be expressed on a suitable basis such as per specific sink organ, or per fresh or dry mass of the sink organ.	$mg \cdot s^{-1}$ or $mol \cdot s^{-1}$
phloem mass flux	Flux of a specified solute through a unit of sieve element cross-sectional area.	$mg \cdot m^{-2} \cdot s^{-1}$
translocation speed (velocity)	Linear distance traveled per unit time by the solution in a file of sieve elements or by a concentration front in a phloem bundle.	$m \cdot s^{-1}$
osmotic-potential gradient	The difference in osmotic potential in a file of sieve elements over a specified distance of phloem path.	$MPa \cdot m^{-1}$
volume flow	Net flux of water entering sieve elements.	$m^3 \cdot m^{-2} \cdot s^{-1}$
phloem pressure gradient	The difference in turgor pressure in a file of sieve elements over a specified distance.	$MPa \cdot m^{-1}$

[a] Unfortunately the two terms are used in opposite ways by different authors. Care must be taken to determine which way a given author chooses to apply these terms.

CONSULTANTS

Susan Dunford
University of Cincinnati
Cincinnati, Ohio

Walter Eschrich
Göttingen, Germany

Donald B. Fisher
Washington State University
Pullman, Washington

R.M. Gifford
CSIRO
Canberra, ACT, Australia

Lim C. Ho
Institute of Horticulture
Littlehampton, West Sussex, England

Colin F. Jenner
University of Adelaide
Glen Osmond, South Australia, Australia

John A. Milburn
University of New England
Armidale, NSW, Australia

Peter E. H. Minchin
The Horticulture and Food Research Institute of New Zealand Ltd.
Lower Hutt, New Zealand

John W. Patrick
University of Newcastle
Newcastle, NSW, Australia

Frank B. Salisbury
Utah State University
Logan, Utah

9

ELECTROMAGNETIC RADIATION

Donald T. Krizek
Climate Stress Laboratory
U.S. Department of Agriculture, ARS
Beltsville, Maryland 20705-2350 U.S.A.

John C. Sager
Biomedical Operations and Research Office (MD-RES)
John F. Kennedy Space Center
Kennedy Space Center, Florida 32899-0001 U.S.A.

An accurate description of the radiation environment used in controlled-environment and other studies is fundamental to plant science. Since formal approval of the SI (*Système International d'Unités*) in 1960 by the *Conférence Générale des Poids et Mesures*, there has been increasing interest among plant scientists in trying to standardize terminology used in describing electromagnetic radiation, which includes the human visually evaluated wavelengths called light. The quantities used to describe and evaluate light (e.g., the candela, lumen, lux), however, are not applicable to plant physiology. The following list of terms, symbols, and units is based on recommendations given in the CIE (*Commission Internationale de l'Éclairage*) International Lighting Vocabulary published in 1987 and in other references attached.

Table 1. Terms, Symbols, and Units Basic to Studies of Radiation [a]**.**

Quantity	*Symbol*	Units
RADIATION [a] *INCIDENT ON A FLAT SURFACE (2 DIMENSIONAL):*		
radiant energy [b]	Q_e	J
radiant exposure	H_e	$J \cdot m^{-2}$
energy flux (irradiance)	E_e	$W \cdot m^{-2}$
spectral energy flux [b] **(spectral irradiance)**	$E_{e\lambda}$	$W \cdot m^{-2} \cdot nm^{-1}$
number of photons [b] **(number of quanta)**	N_p	dimensionless
Avogadro's number (mole) of photons	Q_p	mol

Continued

Table 1. Terms, Symbols, and Units Basic to Studies of Radiation (continued)

Quantity	*Symbol*	Units
photon exposure	H_p	$mol \cdot m^{-2}$
photon flux	E_p	$mol \cdot m^{-2} \cdot s^{-1}$
spectral photon flux [c]	$E_{p\lambda}$	$mol \cdot m^{-2} \cdot s^{-1} \cdot nm^{-1}$
RADIATION ARRIVING AT A POINT (3 DIMENSIONAL):		
The quantities, symbols, and units defined for radiation incident on a flat surface (two dimensional) also apply to radiation incident on a point (three dimensional). However, the term **"fluence"** has been defined as the amount of radiation incident on a spherically shaped receiver. Specifically, fluence is the integral of flux at a point over all directions about the point. In normal conditions, where radiation comes from all directions, one must use a spherical sensor to measure fluence.		
MATERIAL OR REACTION RESPONSE TO RADIATION:		
absorptance, absorption factor (ratio of absorbed to incident radiation)	α	dimensionless
absorbance	A	dimensionless
reflectance (ratio of reflected to incident radiation)	ρ	dimensionless
transmittance	τ	dimensionless

[a] The term **intensity** should not be used to describe radiation falling on a surface or a point. **Intensity** (symbol I) refers to the *source*; e.g., the sun or a lamp.

[b] The same symbol is used for the corresponding energy (e) or photon (p) quantity with the subscript used where confusion might occur.

[c] Spectral data should be shown with the abscissa as a wavelength scale with low values to the left. Discrete responses, such as action or emission spectra should be given in terms of photons.

Table 2. Terms Used in Studies on Special Aspects of Photobiology (Some Relate to the Quantum Theory of Light)

Quantity	*Symbol*	Units
PHOTOSYNTHESIS:		
frequency	f	s^{-1}, Hz
wave number	σ	m^{-1}
wavelength	λ	m (visible spectrum: nm)
fluorescence	F	dimensionless
initial	F_ϕ	
maximum	F_{max}	
variable	$F_v = F_{max} - F_\phi$	
terminal	F_Υ	

Continued

Table 2. Terms Used in Studies on Special Aspects of Photobiology (continued)

Quantity	*Symbol*	Units
quantum yield (ratio of effect to number of photons)	σ	dimensionless
half-peak band-width	$\Delta\lambda_{1/2}$	m (nm)
photosynthetically active radiation (Integral over photosynthetically active wavelengths, 400 to 700 nm)	*PAR*	Specify wavelength interval the first time used (e.g., 400 to 700 nm) (See next two entries for units)
photosynthetic irradiance	*PI*	$W \cdot m^{-2}$
photosynthetic photon flux	*PPF*	$mol \cdot m^{-2} \cdot s^{-1}$ (usually $\mu mol \cdot m^{-2} \cdot s^{-1}$) (sometimes $mol \cdot m^{-2} \cdot d^{-1}$)

PHOTOTROPISM: All terminology is referenced to the basic definitions and units of electromagnetic radiation

PHYTOCHROME (phy):[a]

total phytochrome	$Ptot = Pfr + Pr$	dimensionless
far-red-absorbing form	*Pfr*	dimensionless
red-absorbing form	*Pr*	dimensionless
fraction of phytochrome present in Pfr form with respect to Ptot at photoequilibrium	ϕ	dimensionless
difference in absorbance at two different wavelengths	$\Delta A^{\lambda 2}_{\lambda 1}$	dimensionless
change in difference in absorbance after irradiation with a second actinic source	$\Delta\Delta A^{\lambda 2}_{\lambda 1}$	dimensionless

[a] For a detailed description of current phytochrome nomenclature, the reader is referred to Quail et al. (1994)

REFERENCES

ASAE Engineering Practice: ASAE EP285.7. 1988. Use of SI (Metric) Units. American Society of Agricultural Engineers (ASAE), 2950 Niles Road, St. Joseph, Michigan 49085-9659.

ASAE Engineering Practice: ASAE EP402. 1990. Radiation quantities and units. American Society of Agricultural Engineers (ASAE), 2950 Niles Road, St. Joseph, Michigan 49085-9659.

ASAE Engineering Practice: ASAE EP411.2. 1992. Guidelines for measuring and reporting environmental parameters for plant experiments in growth chambers. American Society of Agricultural Engineers (ASAE), 2950 Niles Road, St. Joseph, Michigan 49085-9659. (See appendix C.)

American Society for Horticultural Science Working Group on Growth Chambers and Controlled Environments. 1980. Guidelines for measuring and reporting the environment for plant studies. HortScience 15(6):719-720.

Commission Internationale de l'Éclairage. 1987. International Lighting Vocabulary. CIE Publ. No. 17.4. Genève, Suisse.

Downs, R.J. 1988. Rules for using the International Systems of Units. HortScience 23(5):811-812.

Holmes, M.G., W.H. Klein and J.C. Sager. 1985. Photons, flux, and some light on philology. HortScience 20(1):29-31.

Krizek, D.T. 1982. Guidelines for measuring and reporting environmental conditions in controlled-environment studies. Physiologia Plantarum 56:231-235.

Krizek, D.T. and J.C. McFarlane. 1983. Controlled-environment guidelines. HortScience 18(5):662-664 and Erratum 19(1):17.

McCree, K.J. 1972. The action spectrum, absorbance, and quantum yield of photosynthesis in crop plants. Agricultural Meteorology 9:191-216.

McFarlane, J.C. 1981. Measurement and reporting guidelines for plant growth chamber environments. Plant Science Bulletin 27(2):9-11.

Mitchell, C.A. and H.C. Dostal. 1977. Light intensity, footcandles and lux are obsolete terms. HortScience 12:437-438.

Monteith, J.L. 1984. Consistency and convenience in the choice of units for agricultural science. Experimental Agriculture 20(2):105-117.

NBS Technical Note 910-2. 1978. Self-Study Manual on Optical Radiation Measurements, Part 1—Concepts. United States Government Printing Office, Washington, DC.

North Central Regional 101 Committee on Growth Chamber Use. 1984. Quality assurance procedures for accuracy in environmental monitoring--Draft proposal. Biotronics 13:43-46.

Quail, P.H., W.R. Briggs, J. Chory, R.P. Hangarter, N.P. Harberd, R.E. Kendrick, M. Koorneef, B. Parks, R.A. Sharrock, E. Schäfer, W.F. Thompson and G.C. Whitelam. 1994. Spotlight on phytochrome nomenclature. Plant Cell 6:468-471.

Salisbury, F.B. and C.W. Ross. 1991. Plant Physiology, Fourth Edition. Wadsworth Publishing Co., Belmont, California. Appendix B: Radiant Energy: Some Definitions, pp. 494-501.

Shibles, R.M. 1976. Committee Report: Terminology pertaining to photosynthesis. Crop Science 16:437-439.

Shropshire, Jr., W. and H. Mohr, editors. 1983. Photomorphogenesis. Encyclopedia of Plant Physiology, V. 16A and 16B. Springer-Verlag, New York.

Smith, H. and M.G. Holmes, editors. 1984. Techniques in Photomorphogenesis. Academic Press, New York.

Spomer, L.A. 1980. Guidelines for measuring and reporting environmental factors in controlled environment facilities. Communications in Soil Science and Plant Analysis 11(12):1203-1208.

Spomer, L.A. 1981. Guidelines for measuring and reporting environmental factors in growth chambers. Agronomy Journal 73(2):376-378.

Thimijan, R.W. and R.D. Heins. 1983. Photometric, radiometric, and quantum light units of measure: a review of procedures for interconversion. HortScience 18(6):818-821.

CONSULTANTS

Steven J. Britz
U.S. Department of Agriculture, ARS
Beltsville, Maryland

Gerald F. Deitzer
University of Maryland
College Park, Maryland

Elisabeth Gantt
University of Maryland
College Park, Maryland

J. Michael Robinson
U.S. Department of Agriculture, ARS
Beltsville, Maryland

Walter Shropshire, Jr.
Omega Laboratory
Timonium, Maryland

Ambler Thompson
U.S. Department of Commerce
Gaithersburg, Maryland

III

PLANT BIOCHEMISTRY AND MOLECULAR BIOLOGY

Much of plant physiology and plant science in general is plant biochemistry. For the most part, traditional plant biochemistry is the same as general biochemistry, but a few special features appear in the tables of Chapter 10. During recent years, plant physiologists have become deeply involved in understanding the biochemistry of plant genetics, a field that is often called *molecular biology*, the topic of Chapter 11.

PRECISION LTM-2
Low Temperature Cooling Module
Power
On
Low Coolant
Off
GTC-2
Genetic Thermal Cycler
Genetic Thermal Cycler
On
Off
Power

10

PLANT BIOCHEMISTRY

Clanton C. Black, Jr.
Biochemistry and Molecular Biology Department
Life Sciences Building
University of Georgia
Athens, Georgia 30602 U.S.A.

The following discussion and tables have been extracted from the *Instructions to Authors*, normally published annually in a January issue of *The Journal of Biological Chemistry* (used by permission), modified somewhat to more closely conform with SI notation and with special reference to the plant sciences.

1. INSTRUCTIONS ON CHEMICAL AND MATHEMATICAL USAGE

A. General. In preparing a manuscript for publication, make references in the text to *simple* chemical compounds by the use of formulas when these can be printed in single horizontal lines of type. Do not use two-dimensional formulas in running text. Center chemical equations, structural formula, and mathematical formulas between successive lines of text. **Prepare such structural formulas and mathematical equations in a form suitable for direct photographic reproduction and include them on a duplicate sheet at the end of the paper.** Similarly, long sequences of amino acids or nucleotides usually reproduce better and will be free from printers' errors if they are printed with a laser printer, drawn in ink, or typewritten by the author. (**Boldfaced print** is best.)

B. Ionic Charge should be designated as a numbered superscript following the chemical symbol; e.g., Mg^{2+}, S^{2-}. The notation Mg(II) is also acceptable.

C. Optically Active Isomers. Names of chiral compounds whose absolute configuration is known may be differentiated by the prefixed R- and S- (see IUPAC (1970) *J. Org. Chem.* **35**, 2849-2867). When the compounds can be correlated sterically with glyceraldehyde, serine, or another standard accepted for a specialized class of compound, SMALL CAPITAL LETTERS D-, L-, and DL may be used for chiral compounds and their racemates. Where the direction of optical rotation is all that can be specified, (+)-, (-)-, and (±)- or *dextro*, *laevo*, and "optically inactive" are used, but in such instances the conditions of measurement must be specified.

D. Isotopically Labeled Compounds. The following guidelines conform to the recommendations adopted by the IUB Committee of Editors of Biochemical Journals (CEBJ). For more detailed instructions consult the IUPAC-CNOC Recommendations on Isotopically Modifed Compounds (1978) *Eur. J. Biochem.* **86**, 9-25.

For most biochemical usage, an isotopically labeled compound is indicated by placing the symbol for the isotope introduced in **square brackets directly attached to the front of the name** (word) or formula as in [^{14}C]urea, [a-^{14}C]leucine, L-[methyl-^{14}C]methionine, [^{3}H]CH_4. If the specific position of the labeling is known, it should be indicated at least the first time the compound is mentioned or in the Materials and Methods section; thereafter, the less specific notation can be used. The following rules govern most situations.

The isotopic prefix precedes that part of the name to which it refers, as in sodium [^{14}C]formate, iodo[^{14}C]acetic acid, 2-acetamido-7-[^{131}I]iodofluorene, fructose, 1,6-[1-^{32}P]bisphosphate, ß-hydroxyl[^{14}C] aspartate, 17ß-[^{3}H]estradiol, *E. coli* [^{3}H]DNA. Terms such as ^{131}I-labeled albumin should not be contracted to [^{131}I]albumin, since native albumin does not contain iodine; however, ^{131}I-albumin and [^{131}I]iodoalbumin are both acceptable.

The symbol indicating the configuration should precede the symbol for the isotope; e.g, D-[^{14}C]glucose; L-[1-^{14}C]leucine; (*R*)-[^{14}C]ethanol.

The same rules apply when the labeled compound is designated by a standard abbreviation or symbol, other than the atomic symbol; e.g., [α-^{32}P]ATP or [^{32}P]CMP (not CM^{32}P).

When isotopes of more than one element are introduced, their symbols are arranged in alphabetical order, for example [3-^{14}C,2,3-^{3}H,^{15}N]serine. Only the symbols ^{2}H and ^{3}H should be used for deuterium and tritium, respectively. When more than one position in a substance is labeled by means of the same isotope and the positions are not indicated, the number of labeled atoms is added as a right-hand subscript, as in [$^{14}C_2$]glycolic acid. The symbol U indicates uniform and G general labeling; e.g., [U-^{14}C]glucose (where the ^{14}C is uniformaly distributed among all six positions) and [G-^{14}C]glucose (where the ^{14}C is distributed among all six positions, but not necessarily uniformly); in the latter case it is often sufficient to write simply [^{14}C]glucose.

When known, the positions of isotopic labeling are indicated by Arabic numerals, Greek letters, or italicized prefixes (as appropriate) placed within the square brackets and before the symbol of the element concerned, to which they are attached by a hyphen; examples are [1-^{2}H]ethanol, [1-^{14}C]alanine, L-[2-^{14}C]leucine (or L-[α-^{14}C]leucine), [*carboxy*-^{14}C]leucine, [Me-^{14}C]isoleucine, [2,3-^{14}C]maleicanhydride, [6,7-^{14}C]xanthopterin, [3,4-^{13}C,^{35}S]methionine, L-[*methyl*-^{14}C]methionine, [1-^{14}C,2-^{13}C]acetaldehyde.

The forms $^{14}CO_2$, $^{32}PO_4$, $^{32}P_i$ are acceptable rather than the more formally correct[1] [^{14}C]O_2 or [^{14}C]CO_2, [^{32}P]P_i, etc. However, the square brackets are not to

[1] According to the IUPAC Recommendations (reference above), a distinction is made between isotopically *substituted* compounds (carrier-free material) where square brackets are not used (e.g. $^{14}CO_2$, Na^{125}I, CH_3-C^2H_2-OH, (^{14}C) carbon dioxide, sodium (^{125}I) iodide, (2-$^{3}H_2$) propanol) and isotopically *labeled* compounds where square brackets are used, either in the formula or in front of the name or formula (e.g. [^{14}C]O_2, [^{3}H]CH_3I, Na[^{125}I], C[^{2}H]$_3$$CH_2$O[^{2}H] and other examples given above and in the IUPAC Recommendations).

be used when the isotopic symbol is attached to a word that is not a specific chemical name, abbreviation, or symbol (e.g., ^{131}I-labeled, ^{3}H-ligands, ^{14}C-steroids, ^{14}C-amino-acids). Note that the abbreviation for Curie is Ci. **The SI unit is the bequerel (Bq).** 1 Bq = 1 disintegration per second or 60 disintegrations per minute. 1 Ci = 37 x 10^9 disintegrations per second = 37 GBq. 1000 dpm = 0.45 nCi = 16.7 Bq. The SI unit is preferred.

E. Spectrophotometric Data. Authors reporting spectrophotometric data must indicate the relation between the symbols used. Although a number of alternatives exist, it is recommended that authors follow the symbols and terminology adopted by IUPAC (1970) *Pure Appl. Chem.* **21**, 1. Beer's law may be stated as: $A = -\log_{10}T = \epsilon lc$, where A is the absorbance; T, the transmittance ($= I/I_o$); ϵ, the molar absorbance coefficient; c, the molar concentration of the absorbing substances; and l, the length of the optical path in centimeters. Under these conditions ϵ has the dimensions $L \cdot mol^{-1} \cdot cm^{-1}$ (*not* $cm^2\ mol^{-1}$). The term **absorbance** is preferred to **optical density**.

If Beer's law is not followed by a particular substance in solution, this should be explicitly stated; even in such cases the substance may be characterized by reporting the absorbance at a specified concentration. When spectrophotometric measurements are made with the use of a radiant energy source that is not confined strictly (as in a line spectrum) to the wavelength or frequency specified, the exact value of ϵ will be somewhat ambiguous; report the spectral characteristics of the source.

F. Molecular Weight and Mass. There are two equivalent expressions that should be distinquished: **molecular weight** (M_r, relative molecular mass) is the ratio of the mass of a molecule to 1/12 of the mass of carbon 12. Hence it is dimensionless. **Molecular mass** (symbol m) in contrast is not a ratio and can be expressed in **daltons** (symbol **Da**) or in **atomic mass units** (symbol **u**). The molecular mass is the mass of one molecule of a substance; it is thus the molar mass (M) divided by Avogadro's number. The dalton is defined as 1/12 of the mass of carbon 12.

It is correct to say either "the molecular mass of X is 10,000 daltons" (or "10 kDa") or "the relative molecular mass (molecular weight) M_r = 10,000)," but it is not correct to express M_r in daltons. One can use expressions such as "the 10 kDa peptide" and "the mass of a ribosome is 2.6 x 10^7 daltons" (or "26 MDa"), even for an entity that is not a definable molecule. Avoid the use of k as a shorthand for 1000 or for kDa (kilodalton).

When presenting estimates of relative molecular mass from gel electrophoresis data, be sure to include the scale used to estimate the molecular mass as one of the *ordinates* on the figure, not just the location of the various standards used.

G. Equilibrium and Velocity Constants. Dissociation constants, association constants, and Michaelis constants should ordinarily be written in terms of concentration; the units should always be clearly indicated at the point where the equilibrium constant is defined and where its value is given.

Values of rate constants should be similarly specified, first-order velocity constants being generally given as s^{-1}. Second-order rate constants are ordinarily given in $M^{-1} \cdot s^{-1}$ (better SI: $L \cdot mol^{-1} \cdot s^{-1}$).

H. Composition of Solutions and Buffers. The composition of all solutions and buffers should be specified in sufficient detail to define the concentration of each species. For ordinary buffers, such as 0.1 $mol \cdot L^{-1}$ sodium acetate, pH 5.0, it will be assumed that the molarity refers to the total concentration of the various species that buffer at the indicated pH and that the concentration of the counterion is sufficient to neutralize the charge of the ionized buffer species. The composition of mixtures should be indicated by the use of a diagonal (/) or a colon. Examples are: chloroform:ether (9:1); 1-butanol/acetic acid/water (75:20:5). Hyphens or dashes are not acceptable for this purpose. Give the complete unabbreviated name and the source (or a reference that gives the complete composition) for all culture media.

2. ABBREVIATIONS AND SYMBOLS

All abbreviations used in a text, except those specifically indicated below (see Tables 1 to 10), should be defined in a single footnote, inserted at the beginning of the paper or immediately after the first such abbreviation. Abbreviations used only in a table or figure may be defined in the legend.

Abbreviations are hindrances to readers in fields other than that of the author, to abstractors, and to scientists in foreign countries. Therefore, their use should be restricted to a minimum. On the other hand, it is sometimes convenient to use abbreviations or symbols for the names of chemical substances, particularly in equations, tables, or figures. A limited use of abbreviations and symbols of specified meaning is therefore accepted. However, clarity is more important than brevity.

For some of the most important biochemical reagents, coenzymes, etc., short abbreviations are universally employed; e.g., ATP, NAD, RNA. The creation of new abbreviations of this kind should be restricted to an absolute minimum. **The following abbreviations are obsolete; do not use them: TCA, PCA, DTE, DOC, DMSO (use Me_2SO), SAM (use AdoMet).**

Do not abbreviate pyridoxal, pyridoxamine, deoxypyridoxine, thiamine, cocarboxylase, pantothenate, folate, pteroylglutamate, trichloroacetic acid, perchloric acid, the tricarboxylic acid cycle, and members thereof. **Most trivial names are sufficiently short that they do not need further shortening.**

Names of enzymes are usually not to be abbreviated except in terms of substrates for which accepted abbreviations exist (exceptions are ATPase, DNase, and RNase). Authors should use the Recommended (trivial) Name given by the IUPAC/IUB Committee on Enzyme Nomenclature in "Enzyme Nomenclature Recommendations (1984)" (1984, Academic Press). Except for very common enzymes, the reaction catalyzed should also be included.

Class names, such as fatty acids, protein, etc., or short terms (poly, furan, folate, etc.) are not to be abbreviated even when an associated term is abbreviated or symbolized (e.g., poly(X), not PX; H_4folate, not THF).

Symbols or abbreviations other than those listed in the IUPAC-IUB Recommendations should be used only in those situations where an objective case may be made for necessity; none should be used when pronouns and similar short terms may replace a long word or phrase. They should always be defined in each paper. Such *ad hoc* abbreviations and symbols should not conflict with recommended symbols and should be introduced only when repeated use is required. If, in exceptional circumstances, symbols or abbreviations are used in an abstract, they should be defined in the abstract as well as in the body of the paper.

When *other abbreviations for chemical compounds* are needed, the maximum use should be made of standard chemical symbols (C, H, O, N, P, S, Na, Cl, etc.), numerical multiples (subscripts 2 and 3, not di or D or T etc., as in Me_2SO, Me_3Si-), and of trivial names and their symbols (e.g., folate, *P*, Me, Pr, Bu, Ph, Ac) (see Tables 3 to 10). These symbols may be combined to represent more complex symbols, such as Tos-Arg-OMe, in which the basic structure (arginine) remains recognizable.

One of the areas of biochemistry for which special *symbols* are essential is that of macromolecules. There are three main series of **symbols for the monomeric units**, those for amino acids, monosacharides, and mononucleosides, of which the amino acid series is the oldest.

The monomeric units of proteins are generally designated by three-letter symbols: a capital followed by two lower case letters. The abbreviations should ordinarily not be used for the free monomers in the text of papers. A standard treatment has been devised for the three groups of macromolecules, built up from these units. For the **amino acid residues** in polypeptides, the residue with the free α-amino group (if one is present) is placed at the left of the sequence as written. Where the sequence of residues is known, the symbols are written in order left to right and joined by short lines (dashes, hyphens). Where the sequence is not known, the symbols are separated by commas, enclosed in parentheses. Example: Ala-Gly-(Met,Pro)-Lys means that the sequence of methionine and proline is unknown.

For the **polysaccarides**, symbols for the sugars are joined by short dashes or arrows to indicate the links between units. The position and nature of the links are shown by numerals and the anomeric symbols α and ß. For example:

Maltose Glc*p*α1-4Glc or Glc*p*α1→4Glc
Lactose Gal*p*ß1-4Glc or Gal*p*ß1→4Glc

The arrow points away from the hemiacetal link. If the dash is used, it is assumed that the hemiacetal link is to the left of it. When it is necessary to indicate furanose or pyranose, the italicized (underlined if italics are not available) letter *f*,or *p* after the saccaride symbol may be used; e.g., Rib*f* for ribofuranose.

Macromolecules composed of repeating sequences may be represented by the prefix 'poly' *or* the subscript *n*, both indicating 'polymer of.' The symbols for the monomeric units of the sequence are enclosed in parentheses. Thus, poly(lys) *or* $(Lys)_n$ is polylysine; poly(Ala-Lys) *or* $(Ala\text{-}Lys)_n$ is a linear polymer consisting of alanine and lysine in regular alternating sequence, and poly(Ala, Lys) is the irregular (random) copolymer of equal amounts of these amino acids. Between poly and the parenthesis there is no intervening space or hyphen. The *n* may be replaced by a definite number, an average (e.g., 10), or a range (e.g., 8 to 12), as appropriate. 'Oligo' may replace 'poly' for short chains. See also the legend to Table 8 regarding nucleoside and mucleotide symbols.

Genetics–A guide to nomenclature in bacterial genetics may be found in Demerec, M., Adelberg, E.A., Clark, A.J., and Hartman, P.E. (1966) *Genetics* **54**,

61-76. Note that the genotypes are *italicized* (underlined if italics are not available); phenotypes are not. (See also Chapter 11 in this volume.)

Genetic designations for various bacteria, bacteriophages, animal viruses, algae, and other materials are listed in "Genetic Maps" (Editor Stephen J. O'Brien). This publication can be purchased from Cold Spring Harbor Laboratory, Fulfillment Department. P.O. Box 100 MM, Cold Spring Harbor, NY 11724. The nomenclature of various bacteria is listed in Bergey, *Manual of Determinative Bacteriology*, 8th Edition, Waverly Press, Baltimore, MD 21202. The nomenclature of transposable elements in prokaryotes can be found in A. Campbell et al. (1979) *Gene* **5**, 197-206, or Szybalski and Szybalski (1979) *Gene* **7**, 217-270.

3. THE TABLES

Table 1. Abbreviations of Units of Measurement and of Physical and Chemical Quantities. These abbreviations may be used without definition. They are not followed by periods. The same form is used in the plural. See Chapter 1 and Section II for more information about most of these and other units. Some are not SI units.

Name	Unit/Symbol
Units of Concentration[a]	
molar (moles/liter)	M[b], $mol \cdot L^{-1}$ (preferred)
millimolar (millimoles/liter)	mM (rather than 10^{-3} M), $mmol \cdot L^{-1}$ (preferred)
micromolar (micromoles/liter)	µM (rather than 10^{-6} M), $\mu mol \cdot L^{-1}$ (preferred)
nanomolar (nanomoles/liter)	nM (not mµM), $nmol \cdot L^{-1}$ (preferred)
picomolar (picomoles/liter)	pM (not µµM), $pmol \cdot L^{-1}$ (preferred)
Other Units	
becquerel[c]	Bq
curie	Ci (not SI)
dalton	Da (not SI)
unified atomic mass unit (this is the SI equivalent of the dalton; its use is preferred; see Chapter 1)	u
equivalent (should be avoided)	eq (not SI)
counts per minute	cpm (not SI, but acceptable)
revolutions per minute	rpm (not SI, but acceptable)
cycles per second (hertz)	Hz
calorie	cal (not SI, use J)
kilocalorie	kcal (not SI, use kJ)
swedberg (10^{-13} s)	S (not SI)

Continued

Table 1. Abbreviations of Units of Measurement (continued)

Name	Unit/Symbol
Physical and Chemical Quantities	
absorbance	A
equilibrium constant	K
Michaelis constant	K_m
relative molecular mass	M_r
retardation factor	R_{F}
average acceleration of gravity at earth's surface[d] (use to report centrifugation, etc.)	g_{n}
specific rotation	$[\alpha]^{\mathrm{t}}$
sedimentation coefficient	s
sedimentation coefficient in water at 20 °C, extrapolated to zero concentration	$s^{\circ}_{20,\mathrm{w}}$
diffusion coefficient	D
Thermodynamic Terms[e]	
Gibbs energy change (formerly ΔF)	ΔG
entropy change	ΔS
enthalpy change	ΔH

[a] Terms such as milligram percent (mg%) should not be used. Mass concentrations should be given as g/kg, mass/volume concentrations as g/L, etc. The liter (preferred symbol L) is accepted for use with the SI (see Chapter 1 in this book).

[b] The letter M is not an abbreviation for mole (mol); it is reserved for molar. Use mM for 10^{-3} M and µM for 10^{-6} M. Avoid designating concentrations as µmol per mL. The designation should, in this case, properly be mM (i.e., millimolar). Maintain consistency in the use of units in situations where they are to be compared (e.g., do not juxtapose 10^{-4} M and 10^{-5} M). As discussed in Chapter 1, second-level discussions of the SI state that *molar* and M should be replaced by the more readily understandable (to nonchemists) $\mathrm{mol \cdot L^{-1}}$, $\mathrm{mmol \cdot L^{-1}}$, etc. This is supported by physical chemists although, as indicated in this table, *The Journal of Biological Chemistry* continues to accept the term molar (M).

[c] 1 becquerel = 1 dps or 60 dpm. 1 Ci = 3.7 x 10^{10} Bq (37 GBq). Becquerel is the preferred term in the International System of Units.

[d] In the SI, the symbol g (note italics) stands for the acceleration caused by gravity *at any location* (e.g., on the moon as well as earth). The subscript n (not italics) in the symbol g_{n} indicates that the symbol stands for the average acceleration caused by gravity at the earth's surface (9.80665 m s^{-2}); that is, g_{n} is a unit. (See Chapter 1.)

[e] For thermodynamic terms see the Recommendations of the Interunion Commission on Biothermodynamics (*J. Biol. Chem.* 251, 6879-6886. 1976)

Table 2. Abbreviations for Semisystematic or Trivial Names. Those abbreviations preceded by an asterisk may be used without definition.

	Abbreviation	Name
*	AMP, ADP, and ATP[a]	Adenosine 5′-mono, di-, and triphosphates
*	cAMP, cGMP etc.	Cyclic AMP (adenosine 3′:5′-monophosphate), etc.
	CMP-NeuAc	Cytidine monophospho *N*-acetylneuraminic acid
*	CoA (or CoASH)	Coenzyme A

Continued

Table 2. Abbreviations for Semisystematic or Trivial Names (continued)

	Abbreviation	Name
*	CoASAc	Acetyl coenzyme A
*	Cm-cellulose	*O*-(Carboxymethyl)cellulose
*	CMP, CDP, and CTP[a]	Cytidine 5′ -mono-, di-, and triphosphates
*	DEAE-cellulose	*O*-(Diethylaminoethyl)-cellulose
*	DNA	Deoxyribonucleic acid or deoxyribonucleate
	DOPA or Dopa	3,4-Dihydroxphenylalanine
*	dTMP, dTDP, and dTTP[a]	Thymidine 5′-mono-, di, and triphosphates
*	EDTA	Ethylenediaminetetraacetate
	EGTA	[Ethylenebis(oxyethylenenitrilo)]tetraacetic acid
*	FAD and $FADH_2$	Flavin-adenine dinucleotide and its fully reduced form
*	FMN	Riboflavin 5′-phosphate
	GDP-Fuc	Guanosine diphosphofucose
	GDP-Man	Guanosine diphosphomannose
	GMP, GDP, and GTP[a]	Guanosine 5′-mono-, di, and triphosphates
*	GSH and GSSG	Glutathione and its disulfide form
*	Hb, HbCO, HbO_2, metHb	Hemoglobin, carbon monoxide hemoglobin, oxyhemoglobin, methemoglobin
	Hepes or HEPES	4-(2-Hydroxyethyl)-1-piperazineethanesulfonic acid
*	IMP, IDP, and ITP[a]	Inosine 5′-mono-, di-, and triphosphates
	Me_2SO	Dimethyl sulfoxide
*	NAD, NAD^+, and NADH	Nicotinamide-adenine dinucleotide and its oxidized and reduced forms
*	NADP, $NADP^+$, and NADPH	Nicotinamide-adenine dinucleotide phosphate and its oxidized and reduced forms
*	P_i	Inorganic phosphate
*	PP_i	Inorganic pyrophosphate
*	RNA	Ribonucleic acid or ribonucleate
	SDS	Sodium dodecyl sulfate
	TEAE-cellulose	*O*-(Triethylaminoethyl)cellulose
*	TMP, TDP, and TTP[a]	Ribosylthymine 5′-mono, di-, and triphosphates
*	Tris	Tris(hydroxymethyl)aminomethane
*	UDP-Gal	Uridine diphosphogalactose
	UDP-GalNAc	Uridine diphospho *N*-acetylgalactosamine
*	UDP-Glc	Uridine diphosphoglucose
	UDP-GlcNAc	Uridine diphospho *N*-acetylglucosamine
	UDP-GlcUA	Uridine diphosphoglucuronic acid
	UDP-Xyl	Uridine diphosphoxylose
*	UMP, UDP, and UTP[a]	Uridine 5′-mono, di-, and triphosphates

[a] The d prefix may be used to represent the corresponding deoxyribonucleoside phosphates; e.g., dADP. The various isomers of adenosine monophosphate may be written 2′-AMP-, 3′-AMP, or 5′-AMP (in case of possible ambiguity). A similar procedure may be applied to other nucleoside or deoxyribonucleoside monophosphates.

Table 3. Miscellaneous Symbols. Most of these abbreviations may be used without definition. Some (e.g., Q, K) should be defined the first time they are used.

Name	Symbol	Name	Symbol
Ferredoxin	Fd	Tocopherolquinone	TQ
Menaquinone	MK	Ubiquinone	Q
Plastoquinone	Q	Circular dichroism	CD
Phosphoric acid residue	P- or -P	Optical rotary disperson	ORD
Phylloquinone	K	Nuclear magnetic resonance	NMR
Pteroic acid (pteroyl-)	Pte	Electron spin resonance	ESR
Pteroylglutamic acid[a]	PteGlu	Electron paramagnetic resonance	EPR
Pyridoxyl-	Pxy-	Infrared spectra	IR spectra
Tocopherol	T	Ultraviolet	UV

[a] Folate and folyl- are not abbreviated.

Table 4. Symbols for Amino Acids. The symbols preceded by an asterisk may be used without definition. The use of the one-letter abbreviations (in parentheses) should be restricted to comparisons of long sequences in tables, lists, or figures, and for such special use as tagging three-dimensional models of proteins. They should not be used in papers where the single-letter system for nucleoside sequences is employed, as in repeating codons. Di(-amino acids) are listed in appendix B of Nomenclature of -Amino Acids, CBN (1975) *Biochemistry* **14**, 449-462.

Name	Symbol	Name	Symbol
Alanine	* Ala (A)	Homoserine lactone	Hse>
3-Aminopropionic acid	ßAla	Hydroxylysine	* Hyl
Arginine	* Arg (R)	Hydroxyproline	* Hyp
Asparagine	* Asn (N)	Isoleucine	* Ile
Aspartic acid	* Asp (D)	(I)Leucine	* Leu (L)
Aspartic acid or asparagine	* Asx (B)	Lysine	* Lys (K)
4-Carboxyglutamic acid	Gla	Methionine	* Met (M)
Cysteine	* Cys (C)	Ornithine	* Orn
Glutamic acid	* Glu (E)	Phenylalanine	* Phe (F)
Glutamine	* Gln (Q)	Proline	* Pro (P)
Glutamic acid or glutamine	* Glx (Z)	5-Pyrrolidone-2-carboxylic acid (pyroglutamic acid; 5-oxoproline)	<Glu
Glycine	* Gly (G)	Serine	* Ser (S)
Half-cystine	* Cys |	Threonine	* Thr (T)
Histidine	* His (H)	Tryptophan	* Trp (W)
Homocysteine	Hcy	Tyrosine	* Tyr (Y)
Homoserine	Hse	Valine	* Val (V)

Table 5. Symbols for Carbohydrates and Organic Acids. Those symbols preceded by an asterisk may be used without definition. Pyranose and furanose forms are designated where necessary by the suffixes *p* and *f*.

Carbohydrate	Symbol
Simple sugars	
Arabinose	Ara
Fructose	* Fru
Fucose	Fuc
Galactose	* Gal
Glucose	* Glc
Mannose	* Man
Rhamnose	Rha
Ribose	* Rib
Xylose	Xyl
Derivatives of various sugars	
N-Acetylglucosamine	* GlcNAc
Glucosamine	* GlcN
2-Deoxyglucose[a]	* dGlc
Glucuronic acid	GlcA
Reductive Pentose Phosphate Cycle	RPPC
6 Phosphogluconic Acid	6PGL
Fructose-1,6-bisphosphate	Fru-1,6-BP or FBP or F1,6-P_2
Fructose-6-phosphate	Fpu-6-P or F6P
Xylulose-5-phosphate	Xylu-5-P or X5P
Sedoheptulose-7-phosphate	Sedoh-7-P or S7P
Sedoheptulose 1,7-bisphosphate	Sedoh-1,7-BP or S1,7-P_2
3-Phosphoglycerate	3-PGA
Glyceraldehyde-3-phosphate	GAP
Dihydroxyacetone phosphate	DHAP
Erythrose-4-phosphate	E-4 P or Ery-4-P
Fructose-2,6-bisphosphate	Fru2,6-Bp or F2,6-P_2
Glucose-1,6-bisphosphate	Glu1,6-Bp or G1,6-P_2
Sialic Acid	Sia
Organic Acids of the Tricarboxylic Acid cycle	
Oxalacetic Acid	OAA
Citric Acid	CIT
α-Ketoglutaric Acid, 2-oxoglutorate	α-KG, 2-OG
Succinic Acid	SUC
Malic Acid	MAL
Fumeric Acid	FUM
Isocitric Acid	ISOCIT
Pyruvic Acid	PYR
Phosphoenolpyruvic Acid	PEP
Cis-Aconitate	cACN

[a] The prefix 'd' indicates a 2-deoxysugar. Isomers may be designated similarly with a positional numerial; e.g., 3-deoxyglucose; 3-dGlc.

Note: In cases where the distinction between *N*-acetyl and *O*-acetyl is important, NeuNAc or NeuOAc are acceptable with definition. Likewise, NeuNGc and NeuOGc are acceptable for the glycolyl analogs.

Table 6. Symbols for Pyrimidine and Purine Bases. These symbols should be defined except those marked with an asterisk.

Base	Symbol	
Adenine	*	Ade
'a base'		Base
Cytosine	*	Cyt
Guanine	*	Gua
Hypoxanthine		Hyp
6-Mercaptopurine (thiohypoxanthine)		Shy
Orotate		Oro
'a purine'		Pur
'a pyrimidine'		Pyr
Thymine	*	Thy
Uracil	*	Ura
Xanthine	*	Xan

Table 7. Symbols for Nucleosides and Nucleotides. Symbols preceded by an asterisk may be used without definition. Two systems are recognized, one using three-letter symbols for the common nucleosides and a capital italic P for the phosphoric residue, the other using single capital letters for the common nucleosides and a lower case p for the phosphoric residue. The three-letter symbols should be used whenever chemical changes involving nucleosides or nucleotides are discussed. The one-letter symbols are intended for the nucleoside residues in sequences or partial sequences only; in these they should always be connected by hyphens (for internal phosphodiester 3′-5′ linkages), and the terminal phosphoric residue should be indicated by the letter p. Codons may be indicated in the text without hyphens or the terminal p's. The 2'-deoxyribonucleosides are indicated by the prefix 'd'. For incompletely specified bases in nucleotide sequences, see (1986) *J. Biol. Chem.* 261, 13-17.

Nucleoside	Symbol	
	Three-letter	One-letter
Adenosine	* Ado	* A
Bromouridine	BrUrd	B
Cytidine	* Cyd	* C
Dihydrouridine		D or hU
Guanosine	* Guo	* G
Inosine	* Ino	* I
6-Mercaptopurine ribonucleoside (6-thioinosine)	Sno	M or MP
'a nucleoside'	Nuc	N
Orotidine	Ord	O
Pseudouridine	* Ψrd	* Ψ or Q[a]
'a purine nucleoside'	Puo	R
'a pyrimidine nucleoside'	Pyd	Y

Continued

Table 7. Symbols for Nucleosides and Nucleotides (continued)

Nucleoside	Symbol Three-letter	Symbol One-letter
Ribosylnicotinamide	Nir	
Ribosylthimine	* Thd	* T
Thiouridine	Srd	S or sU
Thymidine (2′-deoxyribosylthymine)	* dThd	* dT
Uridine	* Urd	* U
Xanthosine	* Xao	* X
Phosphoric residue	*-P*	p or - [b]

[a] For computer work

[b] For internal phosphodiester bonds, use a hyphen.

Table 8. Symbols for Specific Preparations of Nucleic Acids. These symbols may be used without definition.

Name	Symbol
Complementary DNA, RNA	cDNA, cRNA
Heterogeneous nuclear RNA	hnRNA
Messenger RNA	mRNA
Mitochondrial DNA, RNA	mtDNA, mtRNA
Nuclear DNA, RNA	nDNA, nRNA
Ribosomal RNA	rRNA
Transfer RNA	tRNA
Chloroplast plastid or DNA, RNA	ctDNA, pDNA, ctRNA, pRNA

Table 9. Buffers. The Buffer names may be used without definition.

Buffer name	Systematic Description
ACES	2-[(2-amino-2-oxoethyl)amino]ethanesulfonic acid
ADA	[(carba18moylmethyl)imino]diacetic acid
BES	2-[bis(2-hydroxyethyl)amino]ethanesulfonic acid
Bicine	N,N-bis(2-hydroxyethyl)glycine
BisTris	2-[bis(2-hydroxyethyl)amino]-2-(hydroxymethyl)-propane-1,3-diol
CAPS	3-(cyclohexylamino)propanesulfonic acid
CHAPS	3-[(3-cholamidopropyl)dimethylammonio]-1-propanesulfonic acid
CHAPSO	3-[(3-cholamidopropyl)dimethylammonio]-2-hydroxy-1-propanesulfonate
CHES	[2-(N-cyclohexylamino)-ethanesulfonic acid]
CDTA	1,2-cyclohexylenedinitrilotetraacetic acid
EDTA	ethylenediaminetetraacetic acid

Continued

Table 9. Buffers (continued)

Buffer name	Systematic Description
EGTA	[ethylenebis(oxyethylenenitrilo)]tetraacetic acid
EPPS	Acceptable abbreviation for HEPPS (use HEPPS definition).
HEPES	4-(2-hydroxyethyl)-1-piperazineethanesulfonic acid
HEPPS	4-(2-hydroxyethyl)-1-piperazinepropanesulfonic acid
MES	4-morpholineethanesulfonic acid
MOPS	4-morpholinepropanesulfonic acid
PIPES	1,4-piperazinediethanesulfonic acid
TAPS	3-{[2-hydroxy-1,1-bis(hydroxymethyl)ethyl]amino}-1-propanesulfonic acid
TEMED	N,N,N',N'-tetramethylethylenediamine
TES	2-{[2-hydroxy-1,1-bis(hydroxymethyl)ethyl]amino}ethanesulfonic acid
Tricine	N-[2-hydroxy-1,1-bis(hydroxymethyl)ethyl]glycine
Tris	2-amino-2-hydroxymethylpropane-1,3-diol

Table 10. Tentative Rules and Recommendations of International Scientific Unions

Group/Title	May be found in[a]
General	
Abbreviations and symbols for chemical names of special interest in biological chemistry.	(1966) *J. Biol. Chem.* 241, 527-533
Abbreviations and symbols: a compilation (1976).	(1977) *Eur. J. Biochem.* 74, 1-6
Citation of bibliographic references in biochemical journals.	(1973) *J. Biol. Chem.* 248, 7279-7280
Biothermodynamics	
Recommendations for the measurement and presentation of biochemical equilibrium data.	(1976) *J. Biol. Chem.* 251, 6879-6885
Recommendations for the presentation of thermodynamic and related data in biology (1985)	(1985) *Eur. J. Biochem.* 153, 429-434
Labeled compounds	
Isotopically modified compounds[b]	(1978) *Eur. J. Biochem.* 86, 9-25 (1979) *Eur. J. Biochem.* 102, 315-316
Stereochemistry	
Fundamental stereochemistry[c]	(1970) *J. Org. Chem.* 25, 2849-2867
Enzymes	
Enzyme nomenclature. Recommendations (1984)	(1984) Academic Press
The nomenclature of multiple forms of enzymes	(1977) *J. Biol. Chem.* 252, 5939-5941
Catalytic activity	
Units of enzyme activity (1978)	(1979) *Eur. J. Biochem.* 97, 319-320
Symbolism and terminology in enzyme kinetics	(1982) *Eur. J. Biochem.* 128, 281-291
Amino acids, peptides, and proteins	
Nomenclature and symbolism for amino acids and peptides. Recommendations (1983)	(1985) *J. Biol. Chem.* 260, 14-42

Continued

Table 10. Tentative Rules and Recommendations (continued)

Group/Title	May be found in[a]
Abbreviations and symbols for the description of the conformation of polypeptide chains	(1970) *J. Biol. Chem.* 245, 6489-6497
Nomenclature of iron-sulfur proteins	(1979) *Eur. J. Biochem.* 93, 427-430
Corrections	(1979) *Eur. J. Biochem.* 102, 315
Nomenclature of peptide hormones	(1975) *J. Biol. Chem.* 250, 3215-3216
Nomenclature of human immunoglobulins	(1974) *Eur. J. Biochem.* 45, 5-6
Nomenclature of glycoproteins, glycopeptides, and peptidoglycans. Recommendations (1985)	(1987) *J. Biol. Chem.* 262, 13-18
Nomenclature of electron-transfer proteins. Recommendations (1989)	(1991) *Eur. J. Biochem.* 200, 599-611
Carbohydrates	
Tentative rules for carbohydrate nomenclature. Part 1. (1969)	(1972) *J. Bio. Chem.* 247, 613-634
Corrections	(1972) *Eur. J. Biochem.* 25, 4
Conformational nomenclature for five- and six-membered ring forms of monosaccharides and their derivatives (1980)	(1980) *Eur. J. Biochem.* 111, 295-298
Nomenclature of unsaturated monosaccharides (1980)	(1981) *Eur. J. Biochem.* 119, 1-3
Nomenclature of branched-chain monosaccharides (1980)	(1981) *Eur. J. Biochem.* 119, 5-8
Abbreviated terminology of oligosaccharide chains (1980)	(1982) *J. Biol. Chem.* 257, 3347-3351
Polysaccharide nomenclature (1980)	(1982) *J. Biol. Chem.* 257, 3352-3354
Symbols for specifying the conformation of polysaccharide chains	(1983) *Eur. J. Biochem.* 131, 5-7
Lipids	
The nomenclature of lipids. Recommendations (1976)	(1977) *Lipids* 12, 455-468
Nucleotides and nucleic acids	
Abbreviations and symbols for nucleic acids, polynucleotides and their constituents	(1970) *J. Biol. Chem.* 245, 5171-5176
Corrections	(1971) *J. Biol. Chem.* 246, 4894
Abbreviations and symbols for the description of conformations of polynucleotide chains	(1983) *Eur. J. Biochem.* 131, 9-15
Nomenclature for incompletely specified bases in nucleic acid sequences. Recommendations (1984)	(1986) *J. Biol. Chem.* 261, 13-17
Phosphorus	
Nomenclature of phosphorus-containing compounds of biochemical importance. Recommendations (1976)	(1977) *Proc. Natl. Acad. Sci. U.S.A.* 74, 2222-2230
Steroids	
The nomenclature of steroids. Revised tentative rules[d]	(1969) *Biochemistry* 8, 2227-2242
Amendments (1971) and corrections	(1971) *Biochemistry* 10, 4994-4995

Continued

Table 10. Tentative Rules and Recommendations (continued)

Group/Title	May be found in[a]
Quinones	
Nomenclature of quinones with isoprenoid side chains	(1975) *Eur. J. Biochem.* 53, 15-18
Carotenoids	
Tentative rules for the nomenclature of carotenoids	(1972) *J. Biol. Chem.* 247, 2633-2643
Revision[e]	(1975) *Biochemistry* 14, 1803
Cyclitols	
The nomenclature of cyclitols. Recommendations (1973)	(1975) *Eur. J. Biochem.* 57, 1-7
Folic acid	
Nomenclature and symbols for folic acid and related compounds	(1967) *Eur. J. Biochem.* 2, 5-6
Corrinoids	
Nomenclature of corrinoids	(1974) *Biochemistry* 13, 1555-1560
Retinoids	
Nomenclature of retinoids	(1983) *J. Biol. Chem.* 258, 5329-5333
Tetrapyrroles	
The nomenclature of tetrapyrroles	(1980) *Eur. J. Biochem.* 108, 1-30
Tocopherols	
Nomenclature of tocopherols and related compounds (1981)	(1982) *Eur. J. Biochem.* 123, 473-475
Miscellaneous (vitamins)	
Trivial names of miscellaneous compounds of importance in biochemistry	(1966) *J. Biol. Chem.* 241, 2987-2994
Vitamin B6	
Nomenclature for vitamin B6 and related compounds	(1973) *Eur. J. Biochem.* 40, 325-327
Vitamin D	
Nomenclature of vitamin D (1981)	(1982) *Eur. J. Biochem.* 124, 223-227

[a] Most of these documents have also been published in other journals, e.g., *Biochemistry, Biochem. J., Eur. J. Biochem., Boichim. Biophys. Acta, ARch. Biochem. Biophys.*
The second edition of a Compendium of these documents is available from Portland Press Inc., Ashgate Publishing Co., Old Post Rd., Brookfield, VT 05036-9704; or Portland Press, Ltd., P.O. Box 32, Commerce Way, Colchester CO2 8HP, Essex, U.K. The Price is £18.00/U.S. $36.00. Postage is £2.00/U.S. $2.50. A 15% discount is allowable on orders for 10 copies or more to a single address. Payment must accompany the order.

[b] The final version may be found in (1979) *Pure Appl. Chem.* 51, 353-380.

[c] The final version may be found in (1976) *Pure Appl. Chem.* 45, 11-30.

[d] The definitive rules for nomenclature of steroids may be found in (1972) *Pure Appl. Chem.* 31, 285-322.

[e] The definitive rules may be found in (1975) *Pure Appl. Chem.* 41, 407-431.

REFERENCE

[Anonymous]. 1994. Instructions to authors. The Journal of Biological Chemistry 269(1):777-785. ©1994 by The American Society for Biochemistry and Molecular Biology, Inc.

CONSULTANTS

William H. Campbell
Michigan Tech. University
Houghton, Michigan

Jan Miernyk
USDA, ARS
Peoria, Illinois

Jack Preiss
Michigan State University
East Lansing, Michigan

Douglas D. Randall
University of Missouri
Columbia, Missouri

Gregory Schmidt
University of Georgia
Athens, Georgia

11

PLANT MOLECULAR BIOLOGY and GENE DESIGNATIONS

Ellen M. Reardon and Carl A. Price
Waksman Institute
Rutgers University
Piscataway, NJ 08855-0759

1. TERMINOLOGY

The words defined in this section represent terminology common to molecular biology integrated with certain phrases useful in biochemistry, microbiology, and genetics. A few terms are defined within other definitions; these are also printed in **boldfaced** type. Words in *italics* are themselves defined elsewhere although italics may also be used for scientific names and even for emphases.

2D gel Two-dimensional gel; an *electrophoretic* technique based on running a gel under one circumstance, e.g., a pH gradient, rotating the gel 90°, and rerunning the gel under different conditions.

35 S promoter A strong *promoter* from cauliflower mosaic virus (CaMV).

amplified-fragment-length polymorphism (AFLP) A technique used in *genome* mapping.

allele An alternate form of a *gene* at a specific *locus*.

anticodon A *nucleotide* triplet complementary to a *codon*.

ballistic transformation The use of a particle gun to insert foreign DNA into a host.

base, kilobase (b, kb) Adenine (A), guanine (G), cytosine (C), thymine (T), and uracil (U). Bases are elements of DNA and RNA. The lengths of DNA and RNA sequences are measured in *bases* (1000 b = 1 kb).

branch point The point in the replication of a *nucleotide* chain where new nucleotides are added.

cauliflower mosaic virus (CaMV) A DNA virus that infects plants.

centromere Region of the *chromosome* to which the mitotic or meiotic spindle attaches.

chloroamphenicol transacetylase (CAT) A bacterial enzyme encoded by *cat*; serves as a *reporter gene* or *selectable marker*.

chromatin The material of *chromosomes*, consisting of DNA and *histone* proteins.

chromosome A self-replicating module of a *genome* consisting of DNA and proteins. Note: nuclear chromosomes are physically very different from chromosomes of organelles and prokaryotes. Viral chromosomes may be of DNA or RNA.

circular A *polynucleotide* chain, usually of DNA, in the form of a circle; 3′- and 5′-ends may be hydrogen bonded or joined covalently (see *closed circular*).

cis Located on the same strand of DNA.

closed circular A covalently closed circle of DNA.

coding sequence A set of *codons* that encode a protein.

codon A three-*nucleotide* segment of RNA specifying an amino acid or translational *stop* signal.

colony A group of cells, normally on a plate, derived from a single cell.

colony hybridization A method for detecting the presence of a specific *sequence* of DNA among bacterial colonies on a plate.

compatibility The coexistence of two genetic systems in the same cell or organism; e.g., *phage* with a host bacterium or organelles with a nucleus.

complementary DNA (cDNA) A single-stranded DNA copy of an RNA made by reverse transcriptase.

complementation The rescue of function to produce a wild-type phenotype by two separate mutants within the same cell; distinguishable from recombination in which a wild-type gene supplements a mutation.

constitutive Genes that are always expressed; housekeeping genes.

construct A *plasmid* in which genetic or structural elements, such as genes or restriction sites, are introduced through artificial means.

controlling elements Sequences that control the expression of genes by inserting into or retracting from them; **transposable elements**; *transposons*.

cosmid A *plasmid* including a lambda phage cos site that permits packaging of the plasmid DNA.

cryo- Very cold conditions, as in liquid nitrogen.

dalton, kilodalton (Da, kDa, kD) A unit of measurement of molecular mass; kD=1000 daltons, where 1 dalton $=1.661 \times 10^{-24}$ g; equivalent to the *unified atomic mass unit* (u), which is an SI unit defined as 1/12 of the mass of an atom of ^{12}C.

deoxyribonucleic acid (DNA) The macromolecule in which genetic information is stored; the primary genetic material of all cells.

domain A three-dimensional arrangement of amino acids with specific catalytic or binding properties.

downstream **Proximal**; in the 3' direction.

ectopic gene A *gene,* usually one transformed into another species, that is expressed in other than its normal location or stage of development.

editing The addition, removal, or substitution of *nucleotides* to a DNA or RNA to generate or restore correct coding.

electrophoresis The movement of a mixture of macromolecules into a semi-solid, porous medium in response to an electric field to determine its mass/charge ratio (size) or to separate it from other components. This technique is used to separate both proteins and nucleic acids. (A liquid medium can also be used, but today such applications are only for special purposes.)

electroporation The opening of pores in cells by electric shock so as to insert foreign DNA.

enhancer A DNA sequence often but not always distal to the *promoter* that increases promoter activity, and therefore, transcription.

exon The coding portion of an RNA *transcript* of a split gene.

expressed sequence tag (EST) cDNAs for protein genes that are expressed under selected conditions. EST often refers to a partial sequence of the cDNA used to identify sites in the genome that encode a specific gene product.

expression, gene expression The accumulation of specific gene products usually under defined environmental conditions.

gel A semi-solid medium into which proteins or nucleic acids are subjected to *electrophoresis*, to determine their size or to separate them from other components.

gene Although it is impossible to "define" a gene to the mutual satisfaction of biochemists, molecular biologists, and geneticists, in this context a concise meaning might be the sequence of DNA that encodes a *gene product* (an RNA or protein), including all *upstream* and *downstream* sequences involved in the *expression* of the gene.

gene family A set of genes whose sequences differ only slightly that encode identical products; within a single species, a *multigene family.*

gene tagging The addition of a marker to a gene; often by the introduction of a *transposon* or insertion element.

gene transfer *Transformation*; the insertion of foreign DNA into a host conferring a trait not previously inherant in that organism.

genome Sum of all genetic information of an organism, nucleus, organelle, or virus.

β-glucuronidase (GUS) Bacterial enzyme encoded by the *uidA* gene; a *selectable marker* whose product can be directly visualized in many tissues.

heat shock An upward temperature shift inducing stress that results in a quantitative or qualitative alteration in *gene expression.*

heterologous Perceived as different; often used in reference to DNA to mean having dissimilar sequences.

histone A small number of highly conserved proteins that complex with nuclear DNA to form *chromatin.*

homologous Perceived as having a common ancestor; often used in reference to DNA to mean having similar sequences.

hybrid duplex A double-stranded molecule composed of complementary DNA or RNA strands, or single strands of complementary DNA or RNA from different species.

hybridization The mating of complementary DNA and RNA strands to form a *hybrid duplex.*

hydropathy plot A map depicting the hydrophilic and hydrophobic domains of a macromolecule.

immunodetection The identification of a specific polypeptide by use of antibodies.

induction Increased expression of a gene in response to an external factor; e.g., expression of the gene encoding nitrate reductase in response to nitrate; see *repression.*

insertion Introduction of a nucleotide or nucleotide chain into RNA or DNA; may occur naturally, as by *insertion element, transposon,* or *T-DNA.*

insertional mutagenesis Change of genetic information by introduction of a nucleotide chain into a gene; gene is usually rendered inactive.

intron Non-coding portion of an RNA *transcript* that is removed during processing.

iso-electric focusing (IEF) Separation of proteins in a gradient gel to the pH of their isoelectric points.

lariat An intermediate structure formed in an RNA molecule during the excision of certain kinds of *introns.*

lawn The confluence of bacterial colonies on a Petri plate.

library Transformed bacteria, cosmids, YACs, etc., whose *inserts* represent the entire *genome* (genomic library), or *transcripts* (*cDNA* library) of an organism expressed under defined conditions.

ligation The joining of insert and plasmid DNA, usually by the action of T4 ligase.

locus Site on a *chromosome* at which a specific *gene*, or *allele* of that gene, is located. In molecular terms a **locus** can be defined to a single nucleotide, whereas the precision of a locus determined by segregation analysis is limited to about 20 to 2000 kb or 1 centimorgan (see *morgan*).

megaplasmid A *plasmid* of 100 kb or more; e.g., **nif** plasmids, which encode enzymes of nitrogen fixation.

missense DNA Genetic error resulting in a gene product with the wrong sequence.

morgan (M) A unit that expresses the relative distance between genes on a *chromosome*; one M equals a crossover value of 100 %; 1 centimorgan is equivalent to 20 to 2,000 kb in higher plants. Named in honor of Thomas Hunt Morgan.

multigene family A set of genes *within* a species that encode similar or identical products.

mutation Any change in the sequence of DNA in a genome.

nonsense DNA Genetic error that results in premature termination of transcription into RNA.

northern blot The hybridization on a membrane of a specific radiolabeled *probe* with RNA transferred from an electrophoretic separation. Note: *northern blot* is not capitalized; see *Southern blot*.

nucleoid Region of a prokaryote, plastid, or mitochondrion where DNA is concentrated.

nucleotide The basic subunit of DNA and RNA, composed of a base (A, T or U, G, C), a sugar (deoxyribose or ribose), and a phosphate.

nucleus The portion of the cell, often membrane bound, that contains the chromatin.

palindrome Inverted repeats of DNA such that the sequence is the same when read forward or backward; e.g., CAGTTGAC, or in the English language, "Madam, I'm Adam."

phage A virus that infects bacteria.

phenotype The display of characteristics exhibited by an organism resulting from the expression of its *genome* in the existing environment.

plaque Clearing of bacterial *lawn* caused by a virus infection initiated by a single bacteriophage.

plaque-forming unit (PFU) A quantitative measure of the number of viruses descended from a single colony required to clear an area of a bacterial *lawn*.

plasmid DNA that can replicate independently within a bacterial cell; engineered *plasmids* serve as cloning vectors for the insertion of foreign DNA into a host.

plating Growing bacterial colonies, transformed or otherwise, on a Petri plate.

ploidy The numbers of copies of a genome found in a given species; **haploid** has one copy; **diploid**, two; etc.

polyadenylation The post-transcriptional addition of adenylates to the 3′- end of a mRNA molecule; **poly(A)$^+$** and **poly(A)$^-$** refer to the presence or absence of multiple A residues.

polymerase The enzyme by which DNA and RNA are replicated from a *template*.

polymerase chain reaction (PCR) A reaction in which a template of DNA is elongated by cycling of temperatures in a reaction chamber with primer(s) and all possible nucleotides in excess.

polynucleotide A chain of *nucleotides*; DNA or RNA.

post-transcriptional regulation Regulation of gene expression after the synthesis of the primary *transcript*, as at the level of processing or degradation of transcripts.

post-translational regulation Control of gene expression after *translation*, as at the level of processing or degradation of proteins.

premature termination Termination of *transcription* before the end of the *coding sequence* resulting in an incomplete transcript.

primary transcript An RNA synthesized directly from its DNA template prior to processing, polyadenylation, or editing.

primer Short, single-stranded nucleotide segment that initiates the replication of DNA from a template.

probe A DNA, RNA, or protein used to identify or isolate a specific target DNA, RNA, or protein.

promoter The region of DNA that binds RNA polymerase II to initiate transcription; promoters may be constitutive or "turned on" in response to a variety of signals. See *signal transduction*.

protein structure: primary (1° structure) The amino-acid sequence of a protein.

protein structure: secondary (2° structure) The local conformation or folding of the protein's backbone, as into helices or pleated sheets.

protein structure: tertiary (3° structure) The three-dimensional organization of a protein.

protein structure: quarternary (4° structure) The organization of some proteins from multiple polypeptide subunits. **Homopolymeric** proteins are composed of identical subunits; **heteropolymeric** proteins are composed of dissimilar subunits.

pseudogenes A gene rendered non-functional by addition or deletion to its structure; probably related to duplicated genes.

pulsed-field electrophoresis The separation of genomic, or very large DNAs in a semi-solid medium by electric pulses rather than constant current.

random-amplified-polymorphic DNA (RAPD) The amplification of sequences of DNA distributed randomly throughout a genome by PCR technology to analyse genetic structure or relatedness of populations.

reading frame A specific mode of reading codon triplets to produce a functional polypeptide.

receptor Recipient of a signal, such as a hormone, protein, light, etc., that binds to a specific site and initiates a reaction. See *signal transduction*.

recombinant An organism whose genetic makeup is altered by the stable *insertion* of a foreign DNA.

regulation Control of *gene expression.*

reporter gene The coding region of a surrogate gene whose product is easily observed, *e.g.,* CAT, GUS, LacZ, fused to the promoter of a native gene whose product is difficult to detect. Demonstration of the surrogate gene analyzes the native gene: its function, localization, and responses to developmental and environmental *signals.*

repressor Negative regulation of a gene in response to an external factor; e.g., decreased expression of the gene encoding nitrate reductase in response to ammonium; see *induction.*

restriction enzyme An enzyme, usually of bacterial origin, that cuts DNA at specific base pairs, producing **restriction fragments.**

restriction-fragment-length polymorphism (RFLP) Characterization of a genome by comparison of sizes of selected **restriction fragments**.

Ri plasmid Plasmid of *Agrobacterium rhizogenes*; plasmid confers "hairy-root" pathogenicity to *Agrobacterium* species.

ribonucleic acid (RNA) A polynucleotide in which the sugar is ribose rather than deoxyribose; the molecule of which messenger-, ribosomal-, and transfer-RNAs, and some viral genomes are composed.

RNA editing The substitution, addition or removal of bases in a *transcript* to produce a "correct" transcript.

scoreable marker A trait that can be identified in a bacterial lawn.

second messenger A small molecule *within* a cell that relays a chemical message from outside the cell, as cyclic AMP.

selectable marker A trait whose presence or absence enables one to grow colonies or organisms with or without a specific essential component; e.g., antibiotic-resistant mutants in the presence of that antibiotic, or in the absence of an otherwise essential nutrient.

selection Identification of a specific trait useful in singling out a transformant.

selective amplification of microsatellite polymorphic loci (SAMPL). A PCR technique used in genome mapping.

sequence A specific string of bases in a protein or polynucleotide. As a verb, to *sequence* means to determine the sequences of amino acids in a protein or bases in a polynucleotide.

signal sequence, signal peptide A portion, usually the N-terminus, of a protein that is recognized and removed co-translationally by the endoplasmic reticulum (ER) prior to transport of the protein into the lumen of the ER; see *transit peptide.*

signal transduction The process by which the perception of an external signal induces a result: the expression of a gene or set of genes; *e.g.*, the steps between the absorption of light by phytochrome and the synthesis of proteins of the light-harvesting complex.

sodium dodecyl sulfate (SDS) A detergent used to separate polymeric, or apoproteins into their monomeric subunits. See *protein structure: quarternary.*

Southern blot Transfer of DNA that has been separated by electrophoresis to a membrane in order to probe it by hybridization; after E.M. Southern. In recognition of Southern's contribution, subsequent researchers named RNA or protein blots after directions also. See *northern blot, western blot.*

spliceosome A complex of macromolecules that removes *introns* and joins *exons* during RNA processing.

splicing The joining of *exons* into a mature mRNA.

split gene A gene interrupted by an *intron*, or non-coding sequence.

stop codon A *codon* that causes termination of *transcription*; specifically UAA, UAG, and UGA.

subtraction library A library of cDNAs containing only those *transcripts* expressed under a defined condition after elimination of transcripts from differing conditions.

suppression The phenotypic correction of a *mutation* without changing any gene sequences.

T-region of Ti plasmid The portion of a *Ti plasmid* that is transferred into the host genome.

telomere Sequences at both termini of each eukaryotic chromosome that facilitate replication.

temperature-sensitive mutation One in which a wild-type phenotype is expressed at the permissive temperature and a mutant phenotype at a non-permissive (usually higher) temperature.

template The strand of DNA or RNA that serves as the model for its own replication or transcription.

Ti plasmid Plasmid of *Agrobacterium tumefaciens* that confers pathogenicity in crown-gall infection.

trans On the opposite strand of DNA; see *cis.*

transcript An RNA that is copied (**transcribed**) from a gene.

transcription Generation of an RNA copy of a DNA sequence.

transduction Introduction of foreign DNA into a *genome* by a *phage vector.*

transformation The introduction of a gene previously not inherent in the *genome* of an organism.

transient gene expression Expression of a foreign or *ectopic gene* that has not been integrated into a genome.

transit peptide The N-terminus of a protein that is recognized post-translationally and (usually) removed during transport of a protein across a membrane; see *signal peptide*.

translation Formation of a protein on a ribosome according to the instructions in a mRNA.

transon *Exons* encoded separately from one another, with other genes between them or on opposite strands.

transposition The insertion or excision of a *transposon* to or from a genome.

transposon A transposable element usually flanked by inverted repeat sequences; see *controlling elements*.

twintron An *intron* within another intron.

upstream Distal or 5′ to the promoter; opposite the direction of transcription; see *downstream.*

vector *Plasmid*; means of introducing foreign DNA into a host.

western blot Immunodetection of proteins blotted to a membrane. Note: *western blot* is not capitalized; see *Southern blot.*

yeast artificial chromosome (YAC) A *construct* including *centromere*, *telomeres*, an origin of replication and multiple cloning sites for the insertion of foreign DNA.

2. GENE DESIGNATIONS

Any discussion of gene nomenclature must be prefixed with the subject to which the terminology is to be applied: specifically, the traditional genes of genetics and plant breeding, contrasted with genes that have been cloned and sequenced.

A. Nomenclatures of traditional genetics. A specific locus on a chromosome that is associated with a unique phenotype will be referred to here as a genetic system based on segregational analysis. Segregational nomenclatures for individual plant species are distinct, and there is no systematic effort to assign common designations to similar genes in multiple plants. Mutants with similar phenotypes may have no similarity at the molecular level; they can only be distinguished on the basis of their *loci* on chromosomes. Three maize mutants with shortened internodes, for example, might be called *dwarf1, dwarf2,* and *dwarf3.* Years later it might be discovered that *dwarf1* is due to a lesion in the biosynthesis of giberellin, *dwarf2* to a defect in a receptor for gibberellin, and *dwarf3* due to some totally unrelated process. A dwarf mutant in arabidopsis might be due to the same lesion in giberellin biosynthesis, but geneticists can not afford to wait until the biochemical function of the gene product has been determined before designations are assigned to the genes.

B. Nomenclature of sequenced plant genes. The Commission on Plant Gene Nomenclature (CPGN) was founded in 1991 under the auspices of the International Society for Plant Molecular Biology. The goal of the CPGN is to unify nomenclature across the plant kingdom: a gene encoding nitrate reductase in tomato or in arabidopsis or in pea would have the same name. The establishment of plant-wide designations, a unique development in biology, is aided by a long tradition among plant scientists of sharing ideas regardless of the species under study.

Once a gene has been isolated and sequenced, we can make comparisons with genes from other organisms. A common nomenclature for sequenced plant genes began in 1983 with a proposal from Hallick and Bottomley for designations for chloroplast genes, now univerally adopted. In 1988 Lonsdale and Leaver proposed a uniform terminology for mitochondrial genes. CPGN extended the basic principles to all sequenced plant genes (CPGN, 1994).

C. Gene families. The guiding principle of the CPGN nomenclature is that *all genes throughout the plant kingdom that encode the same product are members of the same gene family.* (cf. Fig. 1). The *same* product means that if the sequences of two products are identical, they are obviously the same. But small differences in coding sequences commonly occur within and among species, as in *RbcS* genes, such that the products–the small subunits of ribulose-bisphosphate carboxylase–may be very similar but are not identical. These slightly different small subunits are nonetheless perceived as being "the same" and as belonging to a single, plant-wide family of genes.

There are many instances where genes are classified further on the basis of sequence; *e.g.*, four distinguishable sequences encode functionally identical large subunits of ADPglucose synthase, and four distinguishable sequences encode functionally similar superoxide dismutases. For genes encoding both sets of proteins, coding sequences within a set of genes are 80 to 90 %, but may be only 50 % similar *among* sets, both within and among species.

The requirement for similar coding sequences is not absolute. Proteins displaying activities of 1-aminocyclopropane-1-carboxylate oxidase, for example, are encoded by genes with a variety of coding sequences that do not fall into any discernible natural groupings. It seems reasonable in such cases to assign those genes, at least for the present, to a single family.

D. Gene families and multigene families. Multiple genes that belong to the same plant-wide gene family *within a single species* comprise a *multigene family.* This is consistent with conventional usage. Different members of multigene families are distinguished by *member numbers,* based generally on the chronology of isolation.

E. Gene designations. A specific gene in an individual species of plant is fully defined by four fields:

- Membership in a plant-wide gene family
- Species of plant
- Genome (nuclear, chloroplastic, mitochondrial, viral)
- Membership in a multigene family

The gene symbol identifying memberhip in a plant-wide gene family is usually three letters followed by a number, such as *Adh1* or *Gln2,* but more letters may be used as needed. The letters can be followed by a capital letter instead of a number where dictated by prior usage or by homology with bacterial genes, such as *RbcS* or *TubB*.

Nuclear-encoded genes always start with a capital letter, as in the examples above, while genes encoded in chloroplasts or mitochondria start with a lower-case letter, such as *rbcL* or *cox2.* Thus, the gene encoding the small subunit of ribulose-bisphosphate carboxylase is *RbcS* in higher plants but *rbcS* in chromophytic algae.

The four fields denoting a gene should be stated explicity. An abbreviated format can also be used provided it is defined in advance. The third member of a multigene family in *A. thaliana,* encoding the light-harvesting complex type I LHCII, for example (cf. Fig. 1), could be represented as *Lhcb1;At;3.*

F. Public Databases of Plant Genes. Detailed information on genes in a number of plant species are available through a master database, *PGD,* which is maintained by the National Agricultural Library of the USDA. The species represented in *PGD* include *A. thaliana,* maize and other grains, and soybean, and are being incremented on an almost daily basis. *PGD* also contains the CPGN listings of sequenced plant genes. The electronic address of PGD is:

http://probe.nalusda.gov:8300.html

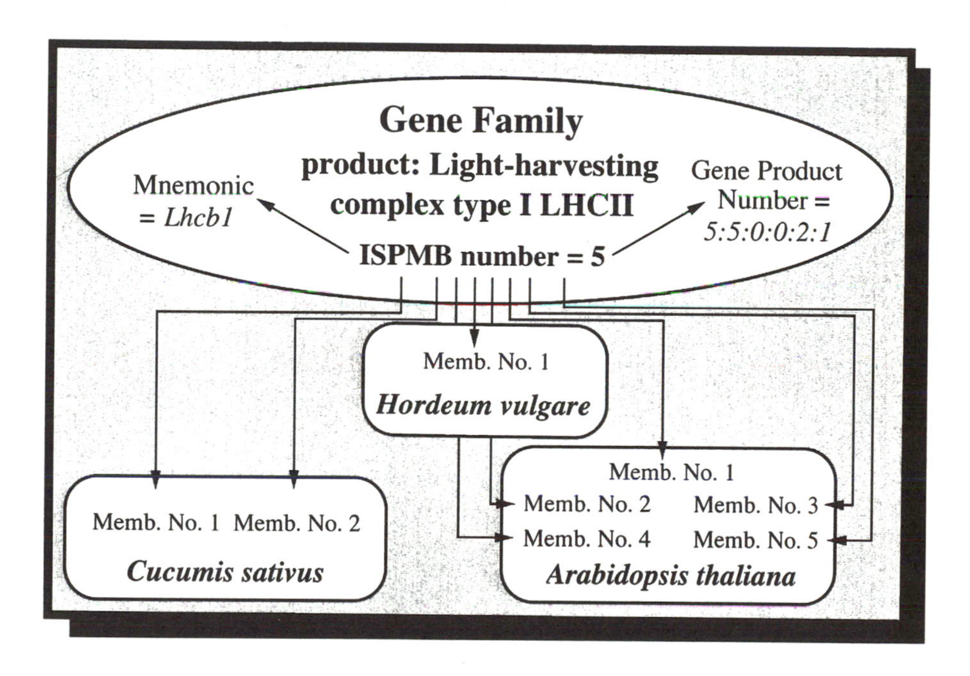

Fig. 1. Designations for plant-wide gene families. The CPGN classifies sequenced plant genes into families based primarily on the function of the gene product. In this example of the light-harvesting complex type I LHCII, every gene in the plant kingdom encoding this protein is a member of this gene family and bears the gene symbol *Lhcb1.* The *ISPMB number* is a consecutively applied identification number used in the management of the CPGN databases. *Gene product numbers* are part of a numerical system being developed by the CPGN for the classification of related families of genes analogous to *Enzyme Commission numbers.* Specific genes in individual species of plants are identified by a name or *mnemonic* of their plant-wide gene family, the genus and species of the plant, and *member numbers,* representative of their occurrence in multigene families.

References

CPGN. 1994. Nomenclature of Sequenced Plant Genes. Plant Mol. Biol. Reptr. 12 Supplement: S1-S109.

Hallick, R.B. 1989. Proposals for the naming of chloroplast genes. II. Update to the nomenclature of genes for thylakoid membrane polypeptides. Plant Mol. Biol. Reptr. 7:266-275.

Hallick, R.B., and W. Bottomley. 1983. Proposals for the naming of chloroplast genes. Plant Mol. Biol. Reptr. 1:38-43.

Jansson, S., E. Pichersky, R. Bassi, B.R. Green, M. Ikeuchi, A. Melis, D.J. Simpson, M. Spangfort, L.A. Staehelin, and J.P. Thornber. 1992. A nomenclature for the genes encoding the chlorophyll *a/b*-binding proteins of higher plants. Plant Mol. Biol. Reptr. 10:242-253.

Lonsdale, D.M. and C.J. Leaver. 1988. Mitochondrial gene nomenclature. Plant Mol. Biol. Reptr. 6(2):14-21.

We are indebted to the authorities who comprise our more than sixty working groups for the development of CPGN's common nomenclature for sequenced plant genes. Without their expertise and continuing input, we would be unable to make databases of approved designations for sequenced plant genes available to the scientific public, and without charge.

Consultants for Section 1

Sewell P. Champe
Waksman Institute
Rutgers University
Piscataway, New Jersey

Eric Lam
Cook College
Rutgers University
New Brunswick, New Jersey

Frank B. Salisbury
Utah State University
Logan, Utah

Julie Vogel
E.I. Du Pont de Nemours & Co., Inc.
Experimental Station
Wilmington, Delaware

IV

PLANT GROWTH AND DEVELOPMENT

The physical and chemical processes in plants that form the basis for the units, symbols, and terms presented in the previous sections all take place in the highly complex machinery that we recognize as a living organism: the roots, stems, and leaves with their xylem and phloem, parenchyma and pith, and all the other tissues made up of cells that in turn are highly complex machines with their cytosol, nucleus, and sundry organelles. Those physical and chemical functions are targets for much study in plant physiology, but the truly astounding thing about living organisms including plants is that they are self generating. All that machinery comes from a single cell, the zygote. Discovering how this happens is the goal of studies in the subfields of plant growth and development. This section attempts to assemble the terms, symbols, and units of measurement used by researchers in those fields.

Most growth and development is influenced, sometimes strongly, by changing environmental factors. This is obvious in the case of the tropisms and the induction of reproductive growth, but it may be a little less obvious in some other cases such as responses to salinity or chilling stress.

This sections begins with terms and units that describe the changes in size and complexity that we recognize as growth and development, continues with plant movements and reproduction, and then presents a large chapter with subsections based on plant responses to various stress factors.

12

MORPHOGENESIS AND THE KINETICS OF PLANT GROWTH

Ralph O. Erickson[1]
Department of Biology
University of Pennsylvania
Philadelphia, Pennsylvania 19104 U.S.A.

This chapter deals with terms used in studying the kinetics of growth. A few terms are defined within other definitions; these are also printed in **boldfaced** type. Words in *italics* are themselves defined elsewhere; italics are also used for symbols that represent physical quantities (as in the rest of this book).

1. THE BIOMETRY OF GROWTH.

absolute growth rate Rate of change of x (size) with respect to t (time), dx/dt; dimensions of $x\ t^{-1}$ (e.g., mm d^{-1}). When the rate is approximately constant for a period of time, growth is then termed **linear**.

allometry Simple allometry obtains when the relative growth rates of two measured attributes of an organ or organism, y and x, are in a constant proportion. Formulated as $y = ax^k$, $\log y = \log a + k \log x$, where a is a constant and k is the allometric coefficient; dimensionless.

anisotropic Having different properties along different directions.

anisotropic growth Magnitude of growth vector differs in different directions.

autocatalytic growth function (or curve) $x = \dfrac{a}{1 + be^{-kt}}$; a symmetrical sigmoid function. The transformation, $\ln(\dfrac{x}{a-x}) = kt - \ln b$, is linear and can be used to evaluate a and k graphically or by least squares for given data on x vs. t.

cell production rate Local rate of formation of new cell walls.

convective rate of change Rate of change associated with movement of a particle to a new location in the growth field.

deposition rate Local net rate of production or import of a metabolite; $d = \dfrac{\partial \rho}{\partial t} + \nabla (\rho \boldsymbol{u})$, where ρ is density, t is time, and $\boldsymbol{u}$ is growth velocity.

---

[1]Current address is: Ralph O. Erickson, 1920 Dog Kennel Road, Media, PA 19063

determinate growth Obtains when the approximate maximum size of an organ or organism is determined, presumedly genetically, as in the growth of a leaf. Growth curve is usually sigmoid.

exponential growth function (or curve) Assuming that the relative rate of change of x with t, $r = d\ln x/dt$, is constant, $x = ae^{rt}$, or $\ln x = \ln a + rt$. Plotting $\ln x$ against t gives a straight line, from which a and r may be estimated.

growth Increase in a measured attribute, x, of an organ or organism, as a function of time, t; $x = f(t)$.

growth field Representation of the spatial distribution of a developmental variable.

growth velocity v_x; rate of displacement of a material particle located at x.

indeterminate growth Obtains when growth is potentially not limited (except by external factors), as is often the case with apical growth of a root or shoot.

kinematics The study of motion and shape change, apart from considerations of mass and force.

local derivative Rate of change associated with a spatial location; e.g., the rate of change of the position 3 mm from the tip of the root.

logistic function Same as *autocatalytic growth function.*

material derivative Rate of change associated with a real, or material, plant element.

monomolecular growth function (or curve) $x = a(1 - e^{-kt})$; represents increase of a measured attribute, x, at a decreasing rate, to a maximum value, a. The linear transformation, $\ln(a - x) = \ln a - kt$, allows graphical or least-squares estimation of a and k from data on x vs. t.

relative elemental growth rate (*RELEL*) For growth along an axis, x, $RELEL = \frac{d}{dx}\left(\frac{dx}{dt}\right)$; for growth of a surface, referred to coordinates, x, y, the growth tensor is used; $REGR = \frac{\partial}{\partial x}\left(\frac{\partial x}{\partial t}\right) + \frac{\partial}{\partial y}\left(\frac{\partial y}{\partial t}\right)$; dimension, t^{-1}.

relative growth rate Rate of change relative to x with respect to time and proportional to x, $\frac{1}{x}\frac{dx}{dt}$, *or* $\frac{d \ln x}{dt}$. May be symbolized r; dimension, t^{-1}. When r is approximately constant for a period of time, growth is then termed **exponential.**

Richards growth function (or curve) $x = a(1 \pm be^{-kt})^{1/(1-m)}$; takes a variety of asymmetrical sigmoid forms. The transformation, $\ln\left(\left(\frac{x}{a}\right)^{1-m} + 1\right) = \ln b - kt$, is linear for appropriate choice of sign, a and m, and can be used to evaluate b and k, from given data on x vs. t. A computer solution is advisable.

sigmoid growth curve A plot of a measured attribute, x, against time, t, showing an early phase of acceleration, and later a deceleration phase, approaching a limiting value.

2. SHOOT AND ROOT MORPHOGENESIS

apical cell The single initial present in the apical meristem of some roots and shoots, as in many lower vascular plants. Divides to form a segment cell **merophyte,** and a **successor apical cell**; this process usually occurring periodically.

decussate Of leaves on a stem, arranged in pairs at each node, each at right angles to the next pair above or below. Symbol, 2(1,1).

development Change from one pattern of growth to another.

differentiation Increase in histological complexity.

divergence angle In *phyllotaxis*, the difference in angular position of two successive leaves on a stem, primordia at the shoot apex, scales on a cone, etc.

Fibonacci sequence The integer sequence, 1, 1, 2, 3, 5, 8, 13,..., in which each term (after the first two) is the sum of the preceding two terms. There are related sequences, such as the **Lucas sequence,** 1, 3, 4, 7, 11,... (See *phyllotaxis*).

generative helix or spiral The sequence of leaves on a stem, or of primordia at the shoot apical meristem, numbered in order of distance along the axis, or in order of radial distance from the center of the meristem. There may be one or more spirals in **k-jugate phyllotaxis.**

growth zone That portion of a root, shoot, or other structure in which cells are formed and enlarge.

initial (1) A meristematic cell that divides to form a new initial cell, plus a cell that divides further to add cells to the plant body; (2) a meristematic element that differentiates into a mature specialized element, as a metaxylem initial.

intrusive growth That type of growth in which the growing cell penetrates between existing cells and in which new areas of contact are formed between the penetrating and neighboring cells (cf. *symplastic growth*).

leaf plastochron index (*LPI*) Indicates the developmental age of a leaf; for a leaf with serial number i, $LPI = PI - i$. When the length of a leaf i, L_i, equals the reference length, L_r, then $LPI = 0$.

meristem That portion of a growth zone in which cell formation (cell division) occurs, accompanied by cell enlargement.

parastichy In *phyllotaxis*, a helical or spiral *rank* of leaves or scales along a stem, or around the center of a shoot apical *meristem*; e.g., a 3-parastichy passes through every third leaf along the stem.

phyllotaxis The arrangement of leaves on a shoot, of primordia at the shoot apical *meristem*, of scales on a pine cone, etc.; formulated by citing a pair of opposed *parastichies* connecting neighboring leaves that differ in serial number by integers, m and n; symbol (m,n). These integers are often consecutive terms of the Fibonacci sequence, as (3,5). Where whorls of k leaves occur at each node, the symbol is $k(m,n)$.

plastochron (1) The period of time between the commencement of two successive, repetitive processes, as between the initiations of two successive leaf primordia; (2) by extension, the time period between corresponding stages of development of two successive leaves.

plastochron ratio (*a*) (1) At the shoot apex, $a = r_n/r_{n+1}$, where r_n is the radial distance from the apex to the center of a leaf primordium n, and r_{n+1}, is the distance to the next younger primordium; (2) more generally, the ratio of a measured attribute of two successive leaves of a growing shoot, as, for leaf length (L), $a = L_n/L_{n+1}$.

plastochron index (*PI*) Indicates comparative developmental age of a shoot:

$$PI = n + \frac{\log L_n - \log L_r}{\log L_n - \log L_{n+1}}$$

, where L is leaf length, L_r is reference leaf length (e.g., L_r = 10 mm), and n is serial number of the leaf just longer than L_r. Assumes leaves are growing exponentially with equal relative rates, and *plastochron* is constant.

relative plastochron growth rate *Relative growth rate* of leaves or leaf primordia, regarding time as measured in plastochrons; $p = \ln a$, where a is the *plastochron ratio*; dimension, t^{-1}, where t is measured in *plastochrons*.

steady growth As in a root, growth in which the pattern of cell production and enlargment is invarient with time.

symplastic growth Adjacent cell walls do not alter their position relative to each other, and no new areas of contact are formed; this leads to expansion with minimum shear in the cross section.

CONSULTANTS

Peter W. Barlow
University of Bristol
Bristol, England

P.W. Gandar
Plant Physiology Division, DSIR
Palmerston North, New Zealand

Paul B. Green
Stanford University
Stanford, California

Zygmunt Hejnowicz
Silesian University
Katowice, Poland

Aristid Lindenmayer
University of Utrecht
Utrecht, Netherlands
(deceased 1989)

R.F. Lyndon
University of Edinburgh
Edinburgh, United Kingdom

T. Sachs
The Hebrew University of Jerusalem
Jerusalem, Israel

Frank B. Salisbury
Utah State University
Logan, Utah

Wendy Kuhn Silk
University of California
Davis, California

R.F. Williams
Australian National University
Canberra City, Australia

13

GROWTH ANALYSIS AND YIELD COMPONENTS

Bruce G. Bugbee
Plants, Soils and Biometeorology Department
Utah State University
Logan, Utah 84322-4820

1. GENERAL CONSIDERATIONS

Plant growth analysis values and yield components can easily and conveniently be expressed with SI units, and yet SI units are often incorrectly used to report these measurements. Like many units reported in the literature, the reasons are historical. A major reference book on the subject of plant growth analysis (Hunt, 1982) used metric and not SI units. However, a recent update of this book (Hunt, 1990) is now consistent with SI rules, except for the use of non-base units in the denominator (mg, weeks), which could be avoided (see Chapter 1). The widespread use of the hectare as a unit of area has caused this unit to remain allowable in many applied agricultural journals, but its use creates a communication barrier between basic and applied researchers that could be resolved if both groups used the SI unit for area: m^2.

The most problematic growth analysis parameters are those with a unit of time in the denominator. Unlike measurements that integrate plant physiological functions over a short time interval (seconds), growth analysis measurements typically integrate the results of many physiological functions over days or weeks. A measured carbon exchange rate would be expressed with SI units as mol $m^{-2}{\cdot}s^{-1}$ with positive values during the light period and negative values during the dark period. The integrated result of the diurnal carbon flux, the crop growth rate, is typically expressed as $kg{\cdot}m^{-2}{\cdot}d^{-1}$. This unit is not strictly allowable according to some authors because of the use of days in the denominator, but the correct unit, $kg{\cdot}m^{-2}{\cdot}s^{-1}$, does not indicate the time of day when the integration occurred. Despite these problems, the recommended time interval is day for crop growth rate. The use of days in the denominator is also important when expressing integrated daily radiation flux (see Chapter 9). Hunt (1990) argues that some biological processes need to be integrated over long intervals and thus uses weeks in the denominator of some growth analysis units. The use of weeks as a unit of time is not recommended. Unlike a 24-h interval, a 7-d interval has no special significance to a plant. When measurements are made at weekly intervals, the results should be reported as a daily average. This avoids the confusion of an additional unit of time.

Specific leaf area, a common growth analysis measurement, has historically been reported as $cm^2\ g^{-1}$. The use of this unit resulted in a range of numbers for the parameter between about 50 and 300 $cm^2\ g^{-1}$, but the gram is not an SI base unit and its use should be avoided in the denominator. The base unit is the kilogram, kg. Presumably in an effort to utilize SI units, specific leaf area has been reported as $m^2\ g^{-1}$, which results in a very small range of values for the parameter (between 0.005 and 0.030). The preferred unit, $m^2\ kg^{-1}$, results in a range of values between 5 and 30; it should therefore be used to express specific leaf area.

2. UNITS FOR GROWTH ANALYSIS AND YIELD COMPONENTS

The following are some examples of preferred units for reporting growth analysis measurements and yield components. Remember that denominators should be SI base units, with the exception of time where day is accepted. Numerators with or without prefixes can be used so that the range of values will be between 0.1 and 1000. The following examples show the base units in both numerators and denominators, as is customary when showing SI units.

Table 1. Terms and Units for Plant Growth Parameters

relative growth rate or specific growth rate	$kg \cdot kg^{-1} \cdot d^{-1}$ (g is appropriate for the numerator; $g \cdot g^{-1} \cdot d^{-1}$ is the unit most often used, but it is not recommended because it uses g, a non-SI base unit in the denominator)
net assimilation rate	$kg \cdot m^{-2} \cdot d^{-1}$ (g is appropriate for the numerator).
leaf area ratio	$m^2 \cdot kg^{-1}$ (m^2 leaf area per kg total dry biomass).
specific leaf area	$m^2 \cdot kg^{-1}$ (m^2 leaf area per kg leaf dry mass).
specific leaf mass	$kg \cdot m^{-2}$ (the inverse of specific leaf area, often called specific leaf weight).
leaf mass ratio	$kg \cdot kg^{-1}$ (kg leaf mass per kg total dry biomass).
crop growth rate	$kg \cdot m^{-2} \cdot d^{-1}$ (g is appropriate for the numerator; it places the typical range of values between 1 and 100).
seasonal crop yield	$kg \cdot m^{-2}$
harvest index	The ratio of harvested or edible (usable) biomass to total biomass or to above-ground biomass. The convention used must be specified. As a ratio, harvest index is dimensionless; it can be expressed as a decimal fraction or as a percent (by multiplying the fraction by 100).
nutrient concentration in plant tissue (N, P, K, etc.)	$mol\ kg^{-1}$. This is typically expressed as a percent, $g \cdot kg^{-1}$, or as ppm. $g \cdot kg^{-1}$ is a correct SI-unit, but inorganic elements are pure substances, and their concentration is thus better expressed as moles rather than as mass. Expressed on a mass basis it appears that N and K are at similar levels in plant tissue; in fact, the number of N atoms is about 3 times higher than the number of K atoms. The physiological importance of nutrients is determined by their number, not their mass.

Other compound units are used in plant growth analysis and yield. Adherence to SI rules should be made with the exception of the use of days in the denominator for some compound units.

Note again that the best unit of area in the denominator is the square meter. If the magnitude of values is above 999 or below 0.1, the value of the unit in the numerator should be changed by adding or deleting the appropriate prefix. For example, 0.03 $kg \cdot m^{-2} \cdot d^{-1}$ is incorrect; 30 $g \cdot m^{-2} \cdot d^{-1}$ is correct. (This rule can be broken if the range of values being compared is greater than 10 000 because the comparison among values is easier if the same prefix is used throughout.)

Table 2. Typical Ranges of Values for Plant Growth Parameters.

Parameter	Unit	Typical Value	Trend over Time
relative growth rate or specific growth rate (*RGR*)	$g \cdot kg^{-1} \cdot d^{-1}$ [a]	10 to 400	decrease
net assimilation rate or unit leaf rate	$g \cdot m^{-2} \cdot d^{-1}$	1 to 30	decrease
leaf area ratio	$m^2 \cdot kg^{-1}$	10 to 60	decrease
specific leaf area	$m^2 \cdot kg^{-1}$	5 to 30	decrease
leaf mass ratio (leaf weight ratio)[b]	$kg \cdot kg^{-1}$	0.4 to 0.8	decrease
crop growth rate	$g \cdot m^{-2} \cdot d^{-1}$	1 to 40	rapid increase then gradual decrease
single leaf photosynthetic rate (CO_2 exchange)	$\mu mol \cdot m^{-2} \cdot s^{-1}$ [c]	1 to 40	decrease
leaf area index	$m^2_{leaf} \cdot m^{-2}_{ground}$	0.01 to 10	large increase (sigmoid curve)

[a] The unit $g \cdot g^{-1} \cdot d^{-1}$ is a commonly used unit for *RGR*, but it is not recommended because the unit of grams is not a base unit and should not be used in the denominator. The unit (d^{-1}) is also used for *RGR* because the grams cancel out. This is confusing, however, because *RGR* represents g new growth per g of existing biomass.

[b] Leaf weight ratio is the commonly used term but kg is a unit of mass, so weight is incorrect.

[c] Use moles for pure substances, such as µmol of CO_2; use kilograms or grams for mixed substances, such as grams of biomass (net assimilation rate or crop growth rate, $g \cdot m^{-2} \cdot d^{-1}$).

Table 3. Growth analysis quantities derived from plant mass and leaf area

	Derived Quantity	Symbol	Instantaneous Value	Formula for Mean Value over Time Interval $(t_2\text{-}t_1)$	Dimensions	Units
Individual Plant	Absolute Growth Rate	*AGR*	$\mathrm{d}m/\mathrm{d}t$	$\mathbf{(m_2 - m_1) / (t_2 - t_1)}$ [a] $\mathbf{= \Delta m/\Delta t}$	$m \cdot t^{-1}$	$\mathrm{kg \cdot d^{-1}}$
	Relative Growth Rate	*RGR*	$1/m \cdot \mathrm{d}m/\mathrm{d}t$	$\mathbf{(\ln m_2 - \ln m_1) / (t_2 - t_1)}$ [b] $\mathbf{= NAR \cdot LAR}$	$m \cdot m^{-1} \cdot t^{-1}$	$\mathrm{kg \cdot kg^{-1} \cdot d^{-1}}$
	Leaf Area Ratio	*LAR*	$\mathbf{L_A / m}$	$[(L_{A1}/m_1) + (L_{A2}/m_2)]/2 = SLA \cdot LMR$	$A \cdot m^{-1}$	$\mathrm{m^2 \cdot kg}$
	Specific Leaf Area	*SLA*	$\mathbf{L_A / L_m}$	$[(L_{A1}/L_{m1}) + (L_{A2}/L_{m2})]/2$	$A \cdot m^{-1}$	$\mathrm{m^2 \cdot kg}$
	Leaf Mass Ratio	*LMR*	$\mathbf{L_m / m}$	$[(L_{m1}/m_1) + (L_{m2}/m_2)]/2$	$m \cdot m^{-1}$	$\mathrm{kg \cdot kg^{-1}}$
	Specific Leaf Mass	*SLM*	$\mathbf{L_m / L_A}$	$[(L_{m1}/L_{A1}) + (L_{m2}/L_{A2})]/2$	$m \cdot A^{-1}$	$\mathrm{kg \cdot m^{-2}}$
	Net Assimilation Rate	*NAR*	$(1/L_A) \cdot \mathrm{d}m/\mathrm{d}t$	$\mathbf{(m_2 - m_1)/(t_2 - t_1) \cdot (\ln L_{A2} - \ln L_{A1})/(L_{A2} - L_{A1})}$	$m \cdot A^{-1} \cdot t^{-1}$	$\mathrm{kg \cdot m^{-2} \cdot d^{-1}}$
	Percent Dry Mass	*PDM*	$\mathbf{(m_D/m_F) \cdot 100}$	$[(m_{D1}/m_{F1}) + (m_{D2}/m_{F2})]/2$	% = 0.01	
Crop	Crop Growth Rate	*CGR*	$1/G_A \cdot \mathrm{d}m/\mathrm{d}t$	$\mathbf{(m_2 - m_1)/(t_2 - t_1) \cdot 1/G_A = NAR \cdot LAI}$	$m \cdot A^{-1} \cdot t^{-1}$	$\mathrm{kg \cdot m^{-2} \cdot d^{-1}}$
	Leaf Area Index	*LAI*	$\mathbf{L_A / G_A}$	$(L_{A1} + L_{A2})/2 \cdot 1/G_A$	$A \cdot A^{-1}$	$\mathrm{m^2 \cdot m^{-2}}$

L_A = Leaf Area, L_m = Leaf mass, G_A = Ground Area, t = time, A = Area, m = mass ≈ Weight, $m_{D \text{ OR } F}$ = Dry or Fresh mass. Note that m = meter, but m = mass (because roman type is used for units but italic type is used for physical quantities; see Chapter 1).

[a] The most frequently used form of each equation is boldfaced.
[b] $\ln \mathrm{m_2} - \ln \mathrm{m_1} = \ln (\mathrm{m_2/m_1})$.

REFERENCES

Causton, David and Jill Venus. 1981. The Biometry of Plant Growth. Edward Arnold, London.
Hunt, R. 1978. Plant Growth Analysis. Edward Arnold, London.
Hunt, R. 1982. Plant Growth Curves: The Functional Approach to Plant Growth Analysis. Univ. Park Press. Baltimore.
Hunt, R. 1990. Basic Growth Analysis. Unwin Hyman Ltd., London, UK; and Winchester, MA.

CONSULTANTS

Ray Wheeler
NASA Kennedy Space Center
Florida

Charles F. Forney
Agriculture Canada
Kentville, Nova Scotia, Canada

Carl Rosen
University of Minnesota
St. Paul, Minnesota

14

PLANT MOVEMENTS

Wolfgang Haupt
Institut für Botanik und Pharmazeutische Biologie
der Universität Erlangen-Nürnberg
Staudtstrasse 5, D-91058 Erlangen GERMANY

This chapter deals with terms used in the study of plant movements. A few terms are defined within other definitions; these are also printed in **boldfaced** type. Words in *italics* are themselves defined elsewhere in this chapter.

1. TYPES AND MECHANISMS OF MOVEMENT

ciliary movements Bending of cilia or flagella of eukaryotic cells, caused by sliding of microtubular doublets along each other, usually resulting in locomotion. The microtubules slide because of a tubulin-dynein interaction.

flagellar movements Rotation of flagella in bacteria, driven by proton-motive force, resulting in locomotion.

running (running phase) Locomotion of **peritrichous bacteria** (having a uniform distribution of flagella over the body surface; movement more or less straight forward), usually caused by counterclockwise rotation of flagella.

tumbling (tumbling phase) Irregular movement of peritrichous bacteria without locomotion, usually caused by clockwise rotation of flagella. The *running phase* is interrupted by a tumbling phase either *autonomously* in more or less regular intervals, or as a response to a *phobic stimulus*. Usually, the direction of movement is randomly different before and after a tumbling phase.

cytoplasmic streaming Movement of cytoplasm (sometimes with nuclei) within the cell as the result of actin-myosin interaction; different types are distinguished according to the regularity of velocity in time and space with which the cell organelles are displaced. (Note that the term **velocity** comprises both speed and direction!)

growth movement Irreversible curvature caused by differential growth; i.e., plastic extension of cell walls, either differentially in a single cell on opposite flanks, or, in a multicellular organ, by differential cell growth on opposite flanks.

Note: A growth curvature can apparently be reversed, but in fact this is caused by an opposite curvature located close to the first one.

bending and bulging Differential growth can cause a curvature in either of two ways: (1) In an organ with a subapical growing zone, curvature is oriented away from the growth-stimulated flank (or toward the growth-inhibited flank): **bending** (sometimes also **bowing**). (2) In an organ with an apical growth center, shifting of maximal growth out of the center results in outgrowth in that direction; hence, curvature is oriented toward the growth-stimulated side: **bulging**.

torsion Growth movement involving twisting of the organ.

turgor movement Reversible curvature caused by asymmetric change in length or volume, either in a single cell by asymmetric elastic extension/relaxation of the cell wall upon uptake or loss of water, or in a multicellular organ by differential swelling/shrinking of cells on opposite flanks.

cohesion movement A change in cell shape resulting in curvature and caused by a loss of water beyond the full relaxation of the extended elastic wall. Because of cohesion of water in the vacuole, its adhesion to the surrounding cytoplasm, and the firm attachment of the latter to the wall, this loss of water results in an elastic inward deformation of the wall (negative turgor). This slow movement can be followed by a sudden return of the cell to its original volume and shape if the elastic deformation becomes stronger than the cohesion, thus allowing water-vapor bubbles to be formed in the vacuole. A classical example: annulus cells of the fern sporangium.

hygroscopic movement (also swelling, shrinking) Reversible curvature caused by water uptake into/water loss out of intermicellary spaces of the cell wall; does not need living structures once a suitable wall texture has been developed.

2. CONTROL OF MOVEMENT: GENERAL

autonomous (endogenous) Self-generated movement, not depending on an external *stimulus*; the controlling factor is sometimes called **internal stimulus**.

induced (exogenous) An external *stimulus* determines that a movement (or a change in movement) starts or how it is executed.

stimulus (signal, input signal) A physical or chemical factor that, when interacting with the organism or cell, can elicit a movement or modify an already existing movement. Its quantity is characterized by its intensity and its duration. It can act in a scalar or in a vectorial way. A stimulus regulates the movement but does not provide the energy for it. Thus, if such a factor also provides energy for the movement, one can debate whether it should be called a stimulus (e.g., photo*kinesis* in cyanobacteria, desmids, etc.).

receptor A structure or substance in cells or organisms, with which a *stimulus* interacts (**stimulus perception**).

perception Interaction between *stimulus* and *receptor* that starts the *transduction* processes.

susception Sometimes the term *perception* is restricted to those processes that involve living structures. In this case, the very first, pure physical effect of the *stimulus* is called *susception*; e.g., displacement (sedimentation) of a statolith-like structure, or excitation of a pigment molecule. *Perception* in the strict sense, then, would be the action of the statolith or the excited pigment molecule on the respective sensitive cell structure (e.g., a membrane).

transduction (stimulus transduction, signal transduction, transduction chain) Biochemical or biophysical events that follow *perception*, finally resulting in the **response**; transduction usually includes amplification processes and a time lag between perception and response.

signal transmission Part of the *transduction* processes bridging the location of *perception* and that of response if they do not occur at the same sites (structures, compartments, cells, organs).

latency (latency time) Time interval between start of *stimulus* and response. Its precise definition depends on the response parameter that is measured (e.g., in photo*tropism* the degree of curvature measured under the microscope or with the naked eye; a differential membrane potential).

reciprocity (reciprocity law; in photoresponses: Bunsen-Roscoe law) The response depends quantitatively neither on the intensity of the *stimulus* alone, nor on its duration, but on the product of both parameters. If reciprocity holds, this is usually taken as indicating a lack of rate-limiting processes in the *transduction* chain or in recovery processes of the *receptor*. Accordingly, reciprocity is restricted to a certain range of intensity and duration of the stimulus. (It often does not hold for very low intensities or very short durations.)

dose That fraction of the applied *stimulus* (product of intensity and duration) that reaches the *receptor*.

adaptation Relaxation of response with persistent or repetitive stimulation; it is sometimes difficult to distinguish whether the *stimulus* changes the *receptor* or *transduction* system such that sensitivity or responsivity is reduced (**sensory** or **response adaptation**).

tonic effect Change in sensitivity or responsivity to a *stimulus* caused by external or internal conditions.

3. TERMS FOR INDUCED MOVEMENTS (TYPES OF RESPONSE)

taxis (tactic orientation; formerly topotaxis) Movement of motile organisms, cells or organelles, the prevailing direction of which is determined by the direction of the *stimulus*. Strictly speaking, taxis is not always a single response, but may be a result of a series of responses (e.g., chemotaxis of bacteria).

phobic response (formerly phobism or phobotaxis) Transient change of movement of motile organisms induced by a change in magnitude of a *stimulus*. The effective change can be an increase or decrease of the magnitude; that is, a *step-up* or *step-down* of the stimulus. After the previous movement is resumed, its direction may have changed, but with no relation to the direction of the stimulus. In inhomogeneous fields, repeated phobic responses can result in patterns of distribution of the organisms; e.g., accumulation in a **"light trap"**.

kinesis The speed of movement of motile organisms or cells depends, under steady-state conditions of input and output, on the magnitude (and sometimes probably also direction) of a *stimulus*.

dinesis Applies when the speed of *cytoplasmic streaming* or the fraction of cytoplasm involved in streaming is controlled by a *stimulus*.

tropism (tropistic movement) Curvature of organs or cells, the direction of which is determined by the direction of the *stimulus*. Tropisms are usually growth movements; **solar tracking**, however, is an example of a **phototropism** that in some plants is caused by turgor changes.

nasty (nastic movement) Curvature of organs or cells, the direction of which is determined by morphological or physiological organization; it is induced by a *stimulus* but independent of its direction. Note: the terms *epinasty, hyponasty, nyctinasty*, *cyclonasty* are used to describe *autonomous* curvatures; thus, they do not deal with true nasties.

growth response Transient change in growth rate induced by a change in magnitude of an external factor. The term is sometimes also used to denote the dependence of steady-state growth rate on a *stimulus*. Mainly used when light is the stimulus; then it is traditionally and inconsistently called **light growth response** rather than **photo growth response.**

strophism *Torsion* induced by an external factor.

4. STIMULI. Note: stimuli are indicated by prefixes attached to -taxis, -nasty, -tropism, etc.:

photo- (formerly: helio-) *Induction* by or orientation with respect to light.

polaro- Orientation with respect to the plane of polarization of radiation.

scoto- *Induction* by darkness.

Note: Strickly speaking, darkness cannot be a *stimulus* but denotes the absence of light as a stimulus.

gravi- (formerly: geo-) Orientation with respect to an acceleration, especially gravity, but also centrifugal acceleration.

agravi- (formerly: ageo-) Indicates that the organ does not respond to the accelerational *stimulus*.

thigmo- (formerly: hapto-) *Induction* by or orientation with respect to mechanical contact with a solid body.

Note: **Thigmotropism** may be used as a general term when a distinction cannot be made between *tropistic* and *nastic* response in tendrils.

seismo- *Induction* by vibration (e.g., by wind action, shaking, usually also by sudden contact with a solid body or a fluid). It should not be used to replace *traumato-*.

thermo- *Induction* by temperature or orientation with respect to a gradient in temperature.

chemo- *Induction* by a specific substance or orientation with respect to a gradient of such a substance.

hygro- (sometimes less correctly: hydro-) *Induction* by fluid water or water vapor, or orientation with respect to a gradient of either.

rheo- *Induction* by or orientation with respect to streaming water or air.

aero- *Induction* by or orientation with respect to air, usually its oxygen content.

traumato- *Induction* by or orientation with respect to wounding.

electro-, galvano- *Induction* by or orientation with respect to an electrical field or current.

magneto- *Induction* by or orientation with respect to a magnetic field.

avoidance response Movement away from a nearby solid barrier.

step-up, step-down Denotes in which direction the magnitude of the *stimulus* has to be changed in order to *induce* a *phobic response* or a *growth response*.

5. DIRECTION OR SENSE OF RESPONSE

positive In scalar stimulation: With increase of *stimulus* the speed of movement increases and vice versa (sometimes also "direct"). In vectorial stimulation: the movement has a prevailing component toward the source of a stimulus.

negative In scalar stimulation: With increase of *stimulus* the speed of movement decreases and vice versa (sometimes also "inverse"). In vectorial stimulation: The movement has a prevailing component away from the stimulus source.

ortho- Orientation parallel or antiparallel to the direction of a vector *stimulus*.

plagio- Orientation at some determined or constant angle between 0° and 180° to the *stimulus* source.

dia- (not transversal) Orientation at right angle to a *stimulus* source.

Note: For a full classification of a movement, one should combine orientation, stimulus, and type of response; e.g., positive phototaxis, plagio-gravitropism.

one-instant mechanism (spatial sensing of direction) The direction of *stimulus* is sensed at one instant by comparing the arriving stimulus at two (or more) *receptor* sites.

two-instant mechanism (temporal sensing of direction) The direction of a *stimulus* is sensed at one *receptor* site only, comparing the arriving stimulus at two instants in time; the arriving stimulus is modulated, e.g., by rotation of the cell or organism during locomotion.

6. TERMS FOR AUTONOMOUS MOVEMENTS. Note: The terms of this section containing *nasty* in the word are not nasties according to the definition; however, they are occasionally used in a purely descriptive way when the controlling factor is not known.

nutation *Autonomous growth movement* over extended periods, i.e., curvature of organs (or change of curvature), caused by differential flank growth, which is not induced by an external *stimulus*. (The term should not be used to denote *nastic* or *tropistic movements*, which are *induced* movements.)

epinasty Growth curvature in a morphologically downward sense, caused by faster growth rate of the upper side (e.g., **adaxial** side of leaves).

hyponasty Growth curvature in a morphologically upward sense, caused by faster growth of the lower side (e.g., **abaxial** side of leaves).

nyctinasty A diurnal periodic upward and downward movement, usually of leaves or petioles, mainly controlled by the physiological clock but, in addition, synchronized by an external factor. Sometimes not restricted to *autonomous* movements, but used also for movements that are *induced* by rhythmic light-dark changes, or only those that are induced by light-dark transition in the sense of *scotonasty*.

circumnutation (formerly: cyclonasty) Periodic change of growth curvature, the tip of the organ ideally moving around a circle or cone; movement occurs as the region of highest growth rate rotates around the organ. Circumnutation is not always *autonomous* but can be the result of *tropistic* stimulations with extended after-effects.

autotropism A tendency of an organ to grow straight and to straighten a curvature induced by a *tropistic stimulus*.

CONSULTANTS

Donat-Peter Häder
Universität Erlangen-Nürnberg
Erlangen, Germany

Anders Johnsson
University of Trondheim
Dragvoll, Norway

Francesco Lenci
C.N.R., Istituto di Biofisica
Pisa, Italy

Hans Machemer
Universität of Bochum
Bochum, Germany

Wilhelm Nultsch
Universität Marburg
Marburg, Germany

Peter Schopfer
Universität Freiburg
Freiburg, Germany

Andreas Sievers
Universität Bonn
Bonn, Germany

Hemming I. Virgin
University of Göteborg
Göteborg, Sweden

Masamitsu Wada
Tokyo Metropolitan University
Tokyo, Japan

Gottfried Wagner
Universität Giessen
Giessen, Germany

Manfred H. Weisenseel
Universität Karlsruhe
Karlsruhe, Germany

15

GROWTH SUBSTANCES

Robert E. Cleland
Department of Botany
University of Washington
Box 355325
Seattle, WA 98195-5325 U.S.A.

Terms used to describe plant growth substances have been used in widely divergent ways. As a result, there has been little attempt to standardize the definitions. The definitions presented here are based, as far as possible, on the most common usage at present. Some terms, such as abscisic acid and ethylene, are not defined here, as they refer to a single compound.

A few terms are defined within other definitions; these are printed in **boldfaced** type. Words in *italics* are themselves defined elsewhere.

antiauxin A compound that antagonizes the biological action of an *auxin*, and whose inhibition kinetics are strictly competitive. For example, p-chlorophenoxyiso-butyric acid is an antiauxin because it shows competitive inhibitor kinetics, but 2,3, 5-triiodobenzoic acid (TIBA) and naphthylphthalamic acid (NPA), which are inhibitors of polar auxin transport and show non-competitive inhibitor kinetics, are not antiauxins.

auxin A compound that has a spectrum of biological activities similar to, but not necessarily identical with those of indoleacetic acid. This includes the ability to: (1) induce cell elongation in isolated coleoptile or stem sections, (2) induce cell division in callus tissues in the presence of a *cytokinin*, (3) promote lateral root formation at the cut surface of stems, (4) induce parthenocarpic tomato fruit growth, and (5) induce ethylene formation.

auxin antagonist A compound that antagonizes the biological action of an *auxin*. The inhibition kinetics can be either competitive or non-competitive. TIBA and NPA can be auxin antagonists, even though they are not *antiauxins*.

bound auxin A molecule in which an *auxin* is bound to another compound (e.g., sugar, amino acid, or macromolecule) via a covalent bond. Sometimes called a **conjugated auxin**.

cytokinin A compound that has a spectrum of biological activities similar to those of trans-zeatin. This includes the ability to: (1) induce cell division in callus cells in the presence of an *auxin*, (2) promote bud or root formation from callus cultures when in appropriate molar ratios to auxin, (3) delay senescence of leaves, and (4) promote expansion of dicot cotyledons. The term cytokinin is often restricted to compounds that contain an adenine ring structure; other compounds with cytokinin activity are called **cytokinin-like**.

florigen A compound or group of compounds, produced in leaves under inductive conditions, that is transported in the phloem to buds and cause their development to change from vegetative to floral. To be considered a florigen, a compound(s) must be effective on a wide range of species, and on short-day, long-day, and day-neutral plants. For this reason, *gibberellins* are not considered to be florigens. As yet, the nature of florigen is unknown.

gibberellin A compound containing the ent-kaurene ring structure. If it is active in a higher plant, it will have a spectrum of biological activities similar to those of gibberellic acid (GA_3). This includes the ability to: (1) promote extension in dwarf genotypes of pea, corn, and rice, (2) induce de novo synthesis of high pI forms of α-amylase in barley aleurone cells, and (3) induce or promote flowering in selected long-day plants when under short-day conditions. **Gibberellin-like** is used for all substances that have GA-like biological activity, but whose chemical structure has not been defined.

hormone-binding protein A protein that binds the hormone in a saturable, specific manner. (See *phytohormone.*)

hormone receptor A *hormone-binding protein* that is shown to have a physiological effect after binding of the hormone.

hormone sensitivity An ill-defined, ambiguous term. In general, it refers to the amount of response (i.e., **responsivity**) to a given amount of hormone. The sensitivity will depend upon: (1) the concentration of hormone from both endogenous and exogenous sources that is present at the site of action, (2) the concentration of hormone receptors, (3) the affinity of the receptors for the hormone, and (4) the coupling of the hormone-receptor to the final observed response. It is not useful to talk about hormone sensitivity or a change in sensitivity unless the term is carefully qualified as to the type of sensitivity that is being considered.

phytohormone (plant hormone) A compound, produced in the plant, that at low concentrations (1 $\mu mol \cdot L^{-1}$ or less) modulates or regulates some aspects of the biochemistry or physiology of cells distant from its site of synthesis. In many cases, hormones may also influence the cells in which they are produced. Examples of phytohormones are: *auxins*, *gibberellins*, *cytokinins*, abscisic acid, ethylene.

plant growth regulator A compound that when applied at low concentrations (1 $\mu mol \cdot L^{-1}$ or less) modifies the growth or development of the plant. A plant growth regulator can be either an endogenous compound (e.g. indoleacetic acid) or a synthetic compound (e.g.α-naphthaleneacetic acid), but the term is primarily applied to synthetic compounds.

polar transport Transport that is dependent on metabolism, more rapid than diffusion, and predominantly unidirectional. The term has primarily been restricted to the movement of hormones, especially *auxins*. Transport through the phloem, where a number of compounds are moving unidirectionally together, is not considered to be polar transport.

CONSULTANTS

Robert Bandurski
Michigan State University
East Lansing, Michigan

Peter J. Davies
Cornell University
Ithaca, New York

Michael L. Evans
Ohio State University
Columbus, Ohio

Richard D. Firn
University of York
York, United Kingdom

Arthur W. Galston
Yale University
New Haven, Connecticut

Russell L. Jones
University of California
Berkeley, California

A. Carl Leopold
Cornell University
Ithaca, New York

Hans Mohr
Universität Freiburg
Freiburg, Germany

Richard P. Pharis
University of Calgary
Calgary, Alberta, Canada

Bernard O. Phinney
University of California
Los Angeles, California

Frank B. Salisbury
Utah State University
Logan, Utah

Andreas Sievers
University of Bonn
Bonn, Germany

Lincoln Taiz
University of California
Santa Cruz, California

Kenneth Thimann
The Quadrangle
Haverford, Pennsylvania

Anthony J. Trewavas
University of Edinburgh
Edinburgh, Scotland

16

BIOLOGICAL TIMING

Willard L. Koukkari
Department of Plant Biology, University of Minnesota
St. Paul, Minnesota 55108 U.S.A.

Beatrice M. Sweeney[1]
Department of Biological Sciences, University of California
Santa Barbara, California 93106 U.S.A.

This chapter deals with terms used in the study of biological timing. A few terms are defined within other definitions; these are also printed in **boldfaced** type. Words in *italics* are themselves defined elsewhere.

acrophase Lag of the maximum (peak) of a mathematical curve (e.g., cosine) versus a reference; the *phase angle* of the crest of the fitted model in relation to a reference time point.

aliasing Misrepresentation of a *frequency* to be lower (or period to be longer) because intervals between consecutively spaced samples were too long. (Sometimes called **folding effect**.)

amplitude Parameter of a rhythm, which for a mathematical (e.g., sinusoidal) curve, is half the **range** from the peak to the trough. It is sometimes (e.g., by certain astronomers and biologists) used for the entire range from peak to trough.

annual Yearly.

biological clock A biological variable showing rhythmicity (especially *circadian rhythms*) and implying a mechanism that imparts time information; see also references to photoperiodism in Chapter 17.

biological cycle Sequence of events in an organism that repeat in the same order and at the same interval through time. [When used in the context of the life history (life cycle) of an organism, the interval through time may vary.] One cycle may be represented as a circle (360°).

---

[1]Deceased.

biological rhythm A change in a biological variable that recurs (repeats) with a specifiable frequency and pattern. Generally viewed as meeting the criteria of an *endogenous rhythm*.

chronogram Plot of a measured variable on the ordinate (Y axis) against the measure of time on the abscissa (X axis) and used to illustrate the state of a variable over time.

circadian pacemaker A postulated master *ocillator* that coordinates the period and phase of *circadian rhythms*.

circadian rhythm *Biological rhythm* having a period of about 24 h (*circa*, about; *dies*, day) and satisfying other criteria, such as having a labile *phase* that can be shifted by external environmental *synchronizers* and continuing for more than one cycle (*free-running*) in the absence of external environmental synchronizers (e.g., under *LL* or *DD*).

circadian time (CT) Time that spans the circadian period. The choice of when a cycle starts is arbitrary, especially under *free-running* conditions. Some biologists designate 00 h as the end of the dark span (**dawn**) of an environmental *LD* cycle, or the phase corresponding to this point in constant conditions (**subjective dawn**). Others select 00 h as the middle of the dark span or the phase corresponding to this point in constant conditions.

circannual rhythm Biological (*endogenous*) rhythm having a period of about (*circa*) a year. Sometimes used when endogenieity has not been demonstrated or is not known.

circaseptan rhythm Biological (*endogenous*) rhythm having a period of about (*circa*) seven days.

circatrigintan rhythm Biological (*endogenous*) rhythm having a period of about (circa) 30 days. (Sometimes called **circalunar rhythm**.)

clock time Time provided by a clock or watch (nonbiological system); also see *biological clock* and *circadian time*.

cosinor analysis Data fitted by cosine(s) of given period(s) (e.g., 24 h and 7 d) by a least-squares procedure and used to estimate rhythm parameters (e.g., *acrophase(s)*, *amplitude(s)*, and *MESOR*).

cosinor (polar) display Summary of rhythm parameters (*amplitude* and *acrophase* of each *harmonic*) estimated by *cosinor analysis* and displayed on polar coordinates so that the phase angle of the maximum (acrophase) of each cosine curve included in the model is shown by the angular direction of a vector and the amplitude of each component is represented by the length of the vector. The 95% confidence limits of both for each harmonic are shown as an ellipse around the tip of the vector. Sometimes the error ellipse is plotted to the rim for purposes of comparing acrophases; in such cases, the length of the vector is not proportional to the amplitude.

damping A decrease in the *amplitude* of an oscillation (rhythm) over time; very common in plants as they mature and develop.

dark break Interruption of the light span or *LL* with a dark span (or pulse).

DD Abbreviation for continuous darkness.

desynchronization A change in the phase relationship between two or more rhythms in an organism (internal); or between the rhythms of the organism and the environmental cycles (external). There may be *desynchronization* in *phase* and/or *frequency*.

diel The 24 h day; 24 h period; rarely, if ever, used when discussing plant rhythms but sometimes used in animal research.

diurnal Term used in reference to either a daily cycle or the light span of a 24 h day (e.g., diurnal animals compared to nocturnal animals). Depending upon the context, often best replaced by other terms (e.g., daily, *circadian*, light span, etc.).

endogenous rhythm A rhythm that persists (free-runs) under constant environmental conditions for more than one complete cycle. To qualify as a *biological rhythm*, the oscillations must be shown to repeat with approximately the same period in the absence of environmental cycles or synchronizers.

entrainment Coupling of the *period* of one rhythm to that of another cycle of about the same length; for example, the setting of a *circadian rhythm* to exactly 24 h by an *LD* environmental cycle. This usually involves both *period* and *phase*.

free-running rhythm See *endogenous rhythm*.

frequency The number of cycles in a unit time or 1/period.

frequency demultiplication Synchronization to a long *period* by a cycle that is a submultiple of that period. (The converse may also be observed, in which case it is referred to as **frequency multiplication**.)

h Abbreviation often used for hour(s). (See Table 5, Chapter 1.)

harmonic Term used to describe features of a periodic curve. A nonsinusoidal periodic function can be mathematically expressed as the sum of cosine curves with period tau (τ), tau/2, tau/3, ...etc. Tau is called the **fundamental period**, and the cosine with period tau is the **fundamental term** or **first harmonic**. Cosines with periods tau/2, tau/3...etc. are called the **second, third**, ... etc. **harmonics**. A curve consisting only of a fundamental term is purely sinusoidal with period tau. [Note: Tau is sometimes used as an abbreviation for *period*.]

high frequency oscillations **Ultradian oscillations** of biological variables having periods less than 30 minutes. See *ultradian rhythms*.

hour-glass timer A mechanism capable of timing only a single time period and likened to an hour-glass (by contrast to a *pendulum timer*); thus, not a rhythm.

infradian rhythm *Biological rhythm* having a *period* appreciably longer than 24 h; usually periods longer than 28 h; therefore, a *circannual rhythm* is infradian.

LD Abbreviation for a light:dark cycle; for example, LD 15:9 would indicate 24 h cycle(s) in which a 15 h light span alternates with a 9 h dark span.

light break Interruption of the dark span or *DD* with a light span (or pulse).

LL Abbreviation for continuous light. Preferably, the energy levels and spectral characteristics should remain constant.

masking Alteration of rhythm parameters or characteristics by external (e.g., environmental) conditions. Masking may be responsible for causing the *amplitude* of a *circadian rhythm* to increase, decrease, or be unexpressed; or it may change a **sinusoidal curve** to a non sinusoidal one or visa versa.

MESOR An acronym (**midline estimating statistic of rhythm**) used in *cosinor analysis* to indicate the mean of the model fitted to the data.

oscillator See *biological clock.*

pendulum timer An oscillating mechanism capable of timing rhythms and likened to the pendulum of a mechanical clock (by contrast to an *hour-glass timer*).

period The time required to complete one cycle. *Biological rhythms* can be classified according to their free-running periods (e.g., *ultradian*, *circadian*, and *infradian*).

phase Measure of timing of any instantaneous state within a cycle (e.g., peak, trough) of a rhythmic variable versus a reference (either internal or external).

phase angle difference Angular difference between the *phase* of one cycle and that of another.

phase response curve A graphic representation of the extent of *phase shifts* caused by treatments (perturbations) of short duration relative to one cycle and positioned at different parts of the cycle (a **pulse experiment**). Phase shifts are often plotted against the rhythm stage (e.g. *circadian time*) when the treatment was administered.

phase shift A change in the *phase* of one cycle relative to that of the original or previous cycle. Sometimes referred to as **rephasing.** If the phase is advanced in time relative to the reference cycle, the phase shift is positive (+) and involves the earlier occurrence of events within a cycle; if the phase is delayed, the phase shift is negative (-) and involves the later occurrence of events within a cycle.

photoperiod Length of the light span in a 24 h *LD* cycle. Commonly used in the literature, although **light span** (or **photofraction**) would be preferable because *photoperiod* has been used to refer to the light span (as defined here) or to the whole daily cycle (comprising both the light and dark spans).

rhythm splitting The subdivision of rhythmic processes into two or more groups with similar or different periods, sometimes observed in organisms under *free-running* conditions.

self-sustained oscillations See *endogenous rhythm.*

semidian rhythm *Biological rhythm* with a period of about 12 h. (The term, **circasemidian,** also appears in the literature to indicate a period of 12 $\pm$ 2 hours.)

sinusoidal rhythm A rhythm that, when the observed measurements are plotted as a function of time, exhibits a curve that approximates a sine curve.

skeleton photoperiod A synchronizing light:dark cycle with short light spans marking the beginning and the end of the usual light span of a 24 h cycle. For example, the skeleton *photoperiod* of a usual *LD* 10:14 cycle could include the following sequence: 30 min light span, 9 h dark span, 30 min light span, and 14 h dark span.

subjective day Under constant conditions (e.g., LL or DD), it is the span corresponding to the light span of a previously used 24 h *LD* environmental cycle.

subjective night Under constant conditions (e.g., LL or DD), it is the span corresponding to the dark span of a previously used 24 h *LD* environmental cycle.

synchronizer An environmental signal that can **rephase** (or **entrain**) a rhythm. Also called the *Zeitgeber*. *Synchronizer* may be preferred since it does not convey the impression of "giving the time" (as Zeitgeber does), but merely to set the phase and/or period of an *endogenous rhythm*.

transient cycle Cycle of an abnormal length occurring immediately after a *phase shift* (or stimulus) for synchronization to a new cycle. More than one transient cycle may be observed before the usual period is reestablished.

ultradian rhythm *Biological rhythm* having a period appreciably less than 24 h (usually less than 20 h). Ultradian oscillations with periods less than 30 minutes are called *high frequency oscillations*.

Zeitgeber See *synchronizer*.

CONSULTANTS

Ruth Satter (deceased)
University of Connecticut
Storrs, Connecticut

Germaine Cornélissen Guillaume
University of Minnesota
Minneapolis, Minnesota

Ola M. Heide
Agricultural University of Norway
AS-NLH, Norway

Bernard Millet
Université de Franche-Comté
Besançon Cedex, France

Frank B. Salisbury
Utah State University
Logan, Utah

Thérèse Vanden Driessche
Université Libre de Bruxelles
Bruxelles, Belgium

17

DORMANCY, PHOTOPERIODISM, AND VERNALIZATION

Frank B. Salisbury
Department of Plants, Soils, and Biometeorology
Utah State University
Logan, UT 84322-4820 U.S.A.

This chapter deals with terms used in the study of vernalization, photoperiodism, and dormancy. A few terms are defined within other definitions; these are also printed in **boldfaced** type. Words in *italics* are themselves defined elsewhere although italics may also be used for scientific names.

abortion Arrest of development of a structure. In some species, especially in bulbous plants, the terms *flower abscission* and *blasting* are frequently used for flower abortion. **Abscission** is the abortion of a structure that shrivels, dries up, and rapidly sheds; **blasting** is the abortion of a structure that shrivels, dries up, but usually does not shed. In certain studies, especially with roses, **blindness** is used for early abortion of the flower.

absolute response See *qualitative response*.

after-ripening Used by some authors with reference to any change that goes on within a dormant seed or bud during the breaking of *dormancy*. Other authors have used the term in a more restricted sense, limiting it to maturation changes that occur in the embryo during storage. The first use is preferable.

allelopathic substances Organic chemicals that are produced by one plant and that harm another plant, sometimes by inhibiting germination.

ambiphotoperiodic plants Plants that respond (e.g., flower) only when given photoperiods that are shorter than some daylength or longer than some longer daylength (e.g., shorter than 14 h or longer than 18 h); opposite to *intermediate-day plants*. This response is rare, but it was reported in *Madia elegans* (Lewis and Went, 1945) and in *Setaria verticillata* (Mathon and Stroun, 1960).

annual A plant with a life cycle from seed to seed that is completed in only one growing season (or one year).

anthesin See *florigen*, which Chailakhyan (1968) suggests is a combination of gibberellin and anthesin.

anthesis The time of coming into full bloom (e.g., in grasses, the time when the anthers are extended from the flower and pollen is released).

antiflorigen A transmissible stimulus that maintains the vegetative state.

apical meristem (apex) A *meristem* borne at the tip of a vegetative plant stem.

autonomously-inductive plant (self-inductive) Flowering occurs more-or-less independently of day length (as in day-neutral plants) and more-or-less independently of any other special environmental treatment. That is, the response occurs under a variety of *constant* environmental conditions.

axillary meristem *Meristems* in the angle (**axil**) formed by the leaf petiole and the stem; potentially capable of forming a branch.

biennial A plant that lives two growing seasons and flowers and dies during the second season. Typically, biennials grow vegetatively during the first season, are induced to flower by the low winter temperature experienced between the seasons, and flower and die the second season; that is, they have an absolute *vernalization* requirement. Often their flowering is also promoted by or requires long days. (Many wild biennials may sometimes be too small after the first season to become vernalized, in which case they might live for more than two seasons, although they flower only once before dying.)

bolting Rapid elongation of a flowering stem from a vegetative *rosette*, often in response to *vernalization* or long days.

caulescent plant A plant with leaves distributed along an elongated stem; opposite to a *rosette plant*.

critical daylength (critical day, critical photoperiod) In plants with an absolute daylength requirement, the daylength or photoperiod that must be exceeded to initiate long-day responses (e.g., flowering of long-day plants) or to inhibit short-day responses. (Some authors have defined critical daylength as the daylength that produces the smallest detectable response, or even 50% flowering, but those definitions should not be used.)

critical nightlength In plants with an absolute daylength requirement, the nightlength or darkperiod that must be exceeded to initiate short-day responses (e.g., flowering of short-day plants or formation of potato tubers) or to inhibit long-day responses. (Some authors have defined critical nightlength as the nightlength that produces the smallest detectable response, or even 50% flowering, but those definitions should not be used.)

daylength (or day length) Sometimes (as in these definitions) written as one word in the sense of the photoperiod in a natural 24-hour cycle of light and darkness, as this might influence plant growth or development. (Otherwise, in English the term is correctly written as two words: *day length*.)

day-neutral plants (DNP) Plants that do not require a specific *daylength* treatment for flower initiation or other photoperiodically controlled response.

determinate With reference to an organ such as a leaf, flower, or fruit that grows to a certain size and then stops growing; stems and roots, because they are produced by *apical meristems*, may continue to grow indefinitely, and are thus *indeterminate*.

developmental arrest A limitation on seed development that prevents a viable embryo from germinating during growth of the seed. (The term could be applied to other structures as well.)

devernalization Reversal of the promotion of flowering induced by exposure to low temperatures (i.e., by *vernalization*) by an immediate exposure to high temperatures (e.g., 30 °C). If a period of time elapses at neutral temperatures between vernalization and the high-temperature treatment, devernalization usually fails. (In some perennial plants, *Chrysanthemum*, for example, prolonged exposure to low irradiance or short days also reverses the effects of vernalization.)

donor In grafting experiments, the graft partner that is assumed to provide the stimulus (promotive or inhibitory) to the *receptor*.

dormancy The condition of a seed or other plant organ when it fails to germinate or grow because it has not been provided with some special set of conditions (e.g., a period of low temperature, suitable wavelengths of light, a treatment that will scarify the seedcoat or leach out inhibitors) although it has been provided with moisture, oxygen, and temperature conditions that are suitable for germination and growth after the special requirements have been met. One special condition can be sufficient time for the embryo to mature.

Dormancy as defined here has been called **endogenous** or **innate dormancy** or **endodormancy** (Lang et al., 1986) as contrasted to **imposed dormancy**, which prevails if an essential factor (e.g., H_2O or O_2) is lacking. Imposed dormancy is called *quiescence* here. **Seed-coat-imposed** dormancy and **embryo dormancy** have also been distinguished.

Pomologists have used *rest* in the sense of *dormancy* as defined here (Samish, 1954). Because this use is rather specialized, it would be well to avoid the term *rest.*

evocation Early responses of receptor tissue following environmental triggering or other form of induction; usually related to the flowering responses that occur at the shoot apex after arrival of flower stimuli and prior to flower differentiation (floral initiation); defined by Evans (1969) to distinguish from *induction*, which occurs in the leaf.

facultative response See *quantitative response.*

florigen (floral stimulus) A postulated flowering hormone or chemical stimulus believed to arise in the leaves of certain plants in response to an appropriate environmental treatment (such as long or short days) and is translocated via the phloem to the bud apices where it causes *evocation* and **flower initiation**. Florigen may also arise autonomously in the leaves of *day-neutral plants* or be transmitted from an induced to a noninduced plant through a graft union.

fractional induction *Induction* caused by one or more cycles of inducing conditions (e.g., short days) interspersed with one or more cycles of noninducing conditions (e.g., long days).

germination The sum of the physical and chemical processes within a seed that lead to visible penetration of the seed coat by the radicle (embryonic root).

grafting The process of combining two separate plants or plant parts (as two stems or a bud and a stem) with the intention that they will unite and grow. The process may be carried out in many ways. See *stock, scion, donor, receptor.*

hard seed Seeds that are kept dormant by hard, impermeable seed coats; can be induced to germinate by physical, chemical, or microbial abrasion/decomposition of the seed coat.

impaction A treatment in which seeds are vigorously shaken to dislodge a cork-like filling (the **strophiolar plug**) in the **strophiolar cleft** of the seedcoat, allowing penetration of water and oxygen and leading to *germination*.

indeterminate The condition of an apical meristem of the shoot or root that has the potential to grow indefinitely. Vegetative *meristems* are indeterminate. Contrasted to a *determinate* meristem, which produces a structure such as a leaf or flower and then ceases to exist as a meristem.

induced state Condition of a plant that has been *induced*.

induction A phenomenon in which some response (e.g., flowering) can be caused (**induced**) in an organism by some treatment (e.g., an environmental condition such as cold or short days), and the response continues, or typically first appears, after the treatment has been discontinued. Sometimes, induction appears to occur *autonomously* in the absence of any obvious treatment (i.e., under a variety of constant environmental conditions). Induction precedes *evocation* and *initiation*.

inhibitor Substance preventing germination, growth, flowering, or other responses; that is, it needs to be leached out, oxidized, or otherwise metabolized (e.g., broken down or bound) to permit the response. (Can also be applied externally to inhibit.)

initiation, flower The beginning of floral differentiation (i.e., a morphological change) that follows *induction* and *evocation*.

intermediate-day plant (IDP) A plant that fails to respond photoperiodically when days are either too short or too long; usually with reference to flower formation.

juvenility State of a usually young plant that is incapable of flowering under otherwise suitable conditions; often associated with a variety of morphological features. (See *ripeness-to-flower*.) Some mature woody plants (i.e., trees) retain juvenile features in their lower parts, a phenomenon sometimes called **secondary juvenility**. Termination of juvenility (achieving *maturity*) may differ from achieving *ripeness-to-flower* in that plants that reach maturity may flower (and/or lose their juvenile morphological features) whether conditions change or not, whereas a plant that has reached ripeness-to- flower typically requires some special treatment (e.g., long days, short days, low temperatures) to actually flower.

long-day plant (LDP) A plant that flowers or otherwise responds when the days are longer than some minimum length (depending on the species) and the nights are shorter than some maximum length; opposite in response to *short-day plant*. Long-day plants typically respond best to continuous light.

long-short-day plant (LSDP) A plant that requires a sequence of long days followed by short days for some photoperiodic response, usually flowering.

maturity The stage reached by a plant when *juvenility* is terminated.

meristem A tissue or group of tissues capable of cell division, enlargement (often elongation), and differentiation to produce *determinate or indeterminate* structures of the root/shoot system. *Apical* or *axillary meristems* of angiosperms may become flowers. Vegetative meristems are *indeterminate*, capable in principle of growing indefinitely to produce stems and roots.

minimum leaf number The minimum number of leaves that must be produced before floral initiation occurs (in plants that produce the first flower at the terminal bud) or that must be produced before a plant achieves the condition known as *ripeness-to-flower* (or *ripeness-to-respond*), in which condition it can become *induced* in response to a photoperiodic or other environmental treatment.

monocarpic species Plants that flower and produce seed only once and then die. They may be *annuals, biennials,* or *perennials*. Contrasted with *polycarpic species*.

night interruption (night break or light break) With reference to inhibition or promotion of some photoperiodic response by interrupting the dark period with an interval of light. Depending upon species, the light break will be more or less effective depending upon when it is given and upon the irradiance level and the spectral quality. *Long-day plants* typically require a much longer period of interruption at a higher irradiance level than do *short-day plants*. If the light break is effective, it acts as a long-day treatment (promoting a long-day response or inhibiting a short-day response). Response to a night interruption is an excellent test for a true photoperiodic response. (See *photoperiodism*.)

nightlength (night length) Sometimes written as one word in the special sense of the dark period in a natural 24-hour cycle of light and darkness, as this might influence plant growth or development. (Otherwise, in English the term is correctly written as two words: night length.)

null-response technique An approach used in studies on photoperiodic phenomena and on other photobiological responses in which irradiance levels of two opposing wavelengths (i.e., red and far-red in most applications) are balanced to produce the same plant response as would be produced by darkness or by some white light source.

perennials Plants that live for an indeterminate number of growing seasons. Most perennials are *polycarpic* and flower once each year when they are sufficiently mature, but some are *monocarpic* (flower only once after several years and then die).

phase change In the context of *juvenility*, the transition from the juvenile to the mature condition, permitting sexual reproduction.

photoperiodism The response of organisms to the relative lengths of day and/or night. Most responses involve changes in growth and development (e.g., flowering, tuber formation, and dormancy). (Adjective form is **photoperiodic**.)

phytochrome Plant pigment consisting of two interconvertible forms absorbing in the red (c.a. 660-nm) or far-red (c.a. 730-nm) regions of light. Phytochrome is directly involved in many photomorphogenic reactions (e.g., photoperiodism, dormancy breaking, deetiolation, chlorophyll formation, etc.). (See Chapter 9.)

polycarpic species Plants that flower for more than one season (often many seasons); all polycarpic species are *perennials*. Contrasted with *monocarpic species*.

prechilling (stratification) The treatment in which imbibed (moist) seeds or dormant plants are subjected to cold temperatures (usually a few degrees above freezing and preferably fluctuating) for some interval of time with the goal of breaking *dormancy* and promoting active growth; not to be confused with *vernalization*, which is a promotion of reproductive growth by cold treatment.

precocious Term applied to switching of developmental pathways prior to natural *maturity*, as in flowering induced chemically or *germination* of immature seed.

qualitative response (absolute response, obligatory response) A plant response, usually flowering, that absolutely depends on some daylength, temperature, or other environmental stimulus. If the plant does not experience the required stimulus, the response does not occur (e.g., the plant remains vegetative). Opposite to *quantitative response*.

quantitative response (facultative response) A plant response that is changed in number or developmental rate (e.g., more flowers or tubers) by a treatment such as a particular day length or an exposure to a period of low temperatures. Opposite to a *qualitative* or *absolute response*. In the absence of the treatment, the response still occurs but at a much slower rate or fewer organs are produced.

quiescence Applied to a *viable*, nondormant seed or bud that fails to germinate or grow only because it has not been provided with suitable temperature, oxygen, and moisture conditions. Has also been called **imposed dormancy** or **ecodormancy** (Lang et al., 1986). See *dormancy*.

recalcitrant seeds Seeds that have a very limited storage period that usually cannot be extended by dry and cold conditions. They are generally large and frequently are tropical species; they germinate at once after ripening.

receptor In grafting experiments, the graft partner that is assumed to receive the stimulus (promotive or inhibitory) from the *donor*.

rejuvenation Reversion (usually only partial) of mature woody plants to a **secondary juvenile phase** by such treatments as pruning, grafting, and treatment with gibberellins.

rest Used by pomologists (Samish, 1954) and others in the same sense as *dormancy* as defined here. *Rest* has also been used in the sense of *quiescence* as defined here. Hence, the term is ambiguous and would best be avoided.

revernalization Effective chilling treatment of plants that have been *devernalized*.

ripeness-to-flower (Blühreife, ripeness-to-respond) A little understood condition of a plant that is reached at a certain age when it is capable of becoming *induced* to flower in response to a photoperiodic or other (e.g., temperature) treatment. The response under discussion is usually that of flowering. See *juvenility, maturity, and phase change.*

rosette plant A plant with leaves coming from a greatly shortened stem at ground level; typical of many biennials during the first year (e.g., beets); opposite to a *caulescent plant.*

scarification Breaking of the seedcoat barrier by mechanical treatment such as abrasion by sand or gravel or microbial action; allows penetration of water and oxygen and thus allows germination.

scion The detached plant part that is grafted to a *stock.*

short-day plant (SDP) A plant that flowers or otherwise responds when the days are shorter than some maximum length (depending on the species) and/or when the nights are longer than some minimum length; opposite to *long-day plant.*

short-long-day plant (SLDP) A plant that responds photoperiodically (flowering is usually the response) to a sequence of short days followed by long days.

stock The root or rooted plant to which a detached plant part (the *scion*) is grafted.

stratification See *prechilling.*

summer dormancy Dormancy of buds established during long days of summer or early autumn, usually in preparation for winter conditions.

thermoperiodism Growth, development, or behavioral responses of organisms to alternating day and night temperatures. In earlier research on storage of bulbs, this term referred to the requirement for different temperatures applied during different developmental stages. To avoid confusion, this second usage should now be avoided.

vernalin A substance postulated to arise in response to *vernalization* and to promote reproductive growth. Its existence is seriously doubted.

vernalization The *induction* of flowering in a plant, moist seed, or developing embryo on the mother plant through exposure to low temperatures (usually a few degrees above freezing). In many (perhaps most?) cases, vernalization leads to *ripeness-to-flower*, after which some other treatment is required to produce flowering. (For example, the biennial strain of *Hyoscyamus niger* has an absolute requirement for long days following vernalization.)

viable Applied to the condition of a seed that is alive and capable of *germination* when provided with suitable environmental conditions; these conditions may include treatments to break *dormancy.*

winter annuals Plants that germinate in late summer or autumn, spending the winter as seedlings, and flowering and fruiting during the next growing season. Flowering may be promoted by exposure to the low temperatures of winter.

REFERENCES

Chailakhyan, Mikhail Kb. 1968. Internal factors of plant flowering. Annual Reviews of Plant Physiology 19:1-36.

Evans, Lloyd T. 1969. The nature of flower induction. In: L.T. Evans, editor. The Induction of Flowering, The Macmillan Company of Australia, South Melbourne. p 457-480.

Lang, Greg, Rebecca Darnell, Jack Early, and George Martin. 1986. Reply to letter. HortScience 21(2):186.

Lewis, Harlan and Frits W. Went. 1945. Plant growth under controlled conditions. IV. Response of California annuals to photoperiod and temperature. Amer. J. Bot. 32:1-12.

Mathon and Stroun. 1960. Third International Congress of Photobiology, Elsevier, Copenhagen. p 384-386.

Samish, R.M. 1954. Dormancy in woody plants. Ann. Rev. of Plant Physiol. 5:183-204.

CONSULTANTS

These terms with preliminary definitions were published in the *Flowering Newsletter*, which was edited and issued by Abraham H. Halevy (now by Georges Bernier). The following scientists responded with comments that strongly influenced the final definition of terms as presented here.

Suresh C. Bhargava
Indian Agricultural Research Institute
New Delhi, India

Georges Bernier
University of Liège
Liège, Belgium

Charles F. Cleland
U.S. Dept. of Agriculture
Washington, D.C.

Abraham H. Halevy
The Hebrew University of Jerusalem
Rehovot, Israel

Wolfgang Haupt
Universität Erlangen–Nürnberg
Erlangen, Germany

Jean-Marie Kinet
University of Liège
Liège, Belgium

Rodney W. King
CSIRO
Black Mountain, Canberra, ACT, Australia

Donald T. Krizek
USDA/ARS
Beltsville, Maryland

Wim de Munk
Bulb Research Center
Lisse, Netherlands

Klaus Napp-Zinn
Botanisches Institut der Universität Köln
(Cologne), Germany

Moshe Negbi
The Hebrew University of Jerusalem
Rehovot, Israel

E. H. Roberts
University of Reading
Reading, England

Kenneth C. Sanderson
Auburn University,
Auburn, Alabama

Max Saure
Diplom-Agraringenieur
Dorfstraße 17
Moisburg, Germany

Walter W. Schwabe
University of London
Wye, England

Atsushi Takimoto
Kyoto University
Kyoto, Japan

Kenneth Thimann
University of California
Santa Cruz, California

Daphne Vince-Prue
Goring-on-Thames
Reading, England

Jan A. D. Zeevaart
Michigan State University
East Lansing, Michigan

18

STRESS PHYSIOLOGY

Leslie H. Fuchigami
Department of Horticulture
Oregon State University
Corvallis, Oregon 97331

Eugene V. Maas
U.S. Salinity Laboratory, USDA ARS
Riverside, California 92507

James M. Lyons
Department of Vegetable Crops
University of California, Davis Campus
Davis, California 95616

D. William Rains
Department of Agronomy and Range Science
University of California, Davis Campus
Davis, California 95616

John K. Raison (deceased)
Plant Physisology Unit
CSIRO Division of Food Research & School of Biological Sciences
Macquarie University
North Ryde, 2113, N.S.W. Australia

Kenneth A. Shackel
Department of Pomology
University of California, Davis Campus
Davis, California 95616-8683

The field of stress physiology is not only of considerable theoretical importance; it is highly significant to agriculture. In practice, researchers tend to specialize within at least four subfields: chilling injury, cold stress, water stress, and salinity stress. Yet, these subfields have several basic terms in common. Thus, this chapter begins with the terms common to study of all plant stresses and is then divided into four sections representing the four subfields. Some **bold-face** terms are also defined in the context of other definitions; words in *italics* are defined elsewhere or are botanical names. See Chapter 6 for terms, units, and symbols used to describe plant water relations (e.g., *water potential and osmotic, matric, and pressure potentials, etc.).*

1. GENERAL STRESS-PHYSIOLOGY TERMS

acclimation Adjustment of an organism to changes in external environments; these anatomical or physiological changes are beneficial and increase the organism's resistance or tolerance to subsequent environmental stress.

adaptation Anatomical or physiological characteristics of an organism, usually genetically fixed, that enable it to live in a given environment.

avoidance An *acclimation* or *adaptation* that reduces the intensity of stress at the cellular level.

calorie The unit of energy required to raise the temperature of one gram of water by one degree centigrade. Equal to **4.1842 joules** (exactly), the preferred SI unit.

conditioning Exposure to temperatures slightly above the *critical temperature* chilling range for various periods or other treatments that can limit the magnitude or affect or delay the onset of the *primary* and/or *secondary events* leading to the development of visible symptoms.

critical temperature The lowest ambient temperature at which the whole or parts of a living organism can endure for 30 min without injury. The critical (**threshold**) temperature may vary with the species, tissue, stage of growth, etc. Some species are damaged by cool temperatures above the freezing point (*chilling-sensitive*); others only by subfreezing temperatures. (Critical temperature is not to be confused with the chemical definition of critical temperature, which is the temperature of vaporization of a liquid.)

dehardening Synonymous with **deacclimation**. The loss in plant tissues and organs of resistance to various *stresses*.

hardening *Conditioning* or *acclimation* of an organism to a particular *stress*, which results in increased resistance to that stress and sometimes to other stresses. With relation to low-temperature stresses, *hardening* is a term used to describe physiological events that lead to a lowering of the *critical temperature* for lethal injury of chilling-insensitive plants exposed to freezing temperatures. The term *hardening* should not be used in relation to chilling; the more appropriate term is *conditioning.*

photo oxidation Oxidation of a substance caused by the absorption of a photon.

primary event(s) The primary cellular sensor(s) or trigger(s) that initiate(s) a series of *secondary events* leading eventually to the visible symptoms of such *strains* as *chilling injury*. The primary event(s) must occur at the critical stress level (e.g., *critical temperature*) for the species or tissue, must be rapid, and in the short term, reversible.

regrowth The ultimate *viability* test to determine survival following a stress. Plants or cells are grown or quieted for a given period of time following a stress and then evaluated for either root and shoot regeneration or increase in mass.

repair The process by which the stress-induced injury of a plant is partially or completely reversed following removal of the stress.

secondary event(s) The metabolic and cellular changes following (directly or indirectly) as consequences of the *primary event* and that lead to the visible symptoms of stress-induced injury. The secondary events are both time- and stress-dependent. In the short term the changes induced are reversible if the stress is removed. However, if the stress is maintained, the tissue becomes unable to recover.

strain The observed (deleterious) biological changes that occur in response to *stress*. (The terms *stress* and *strain* can be used in a manner analogous to their use in physics: *stress* is the force applied to an object, as a metal bar; *strain* is the actual change in shape of the object, as bending of the bar.)

stress Any environmental condition that is capable of causing a biologically injurious change (*strain*). Since plants are autotrophs, any change that directly or indirectly reduces **plant growth** (biomass accumulation) must be considered biologically injurious, even if there are beneficial consequences for other aspects of plant function. Complete descriptions of a given *stress* (or strain) should include its magnitude, duration, and rate of development.

symptoms of injury Visible manifestations of the *secondary events* that reflect injury to the cells and tissues caused by chilling.

tolerance Ability of an organism or its cells or other parts to survive the full impact of a stress. See also *avoidance*.

viability The state of living, growing, or developing. The **viability test** estimates the relative or absolute survival of an organism. For example, common cold-hardiness viability tests include visual browning, conductivity, regrowth, vital stains such as 2,3,5-triphenyl tetrazolium chloride, florescein diacetate, Evan's blue, and neutral red.

2. CHILLING INJURY[1]

ameliorate To provide a treatment or set of conditions that reduce the impact of a *chilling treatment* by altering the time course of symptom development. Thus, amelioration is confined to describing changes in the *tolerance* of the plant tissue to the imposed chilling *stress*.

chilling The act of exposing plant material to a non-freezing low temperature. This exposure may or may not be beneficial to the plant.

chilling injury A descriptive term for the physiological injury to many plants, particularly those warm-season species (e.g., crops) of tropical or sub-tropical origin, when they are exposed to low, but non-freezing temperatures.

chilling-insensitive Those plants that typically continue to grow and develop, albeit slowly, and can complete their life cycle when continuously exposed to chilling temperatures. These plants are primarily cool season species of temperate origin.

---

[1]Original authors of this section were J.K. Raison and J.M. Lyons.

chilling repair Cellular changes that correct the adverse effects of the *secondary events*.

chilling reversal Reversal of the *primary event(s)*, which would be rapid and direct and shut off further stimulation of the secondary events.

chilling-sensitive plants Those plants that are injured by exposure to temperatures below about 10 °C to 15 °C, but above freezing. (The warm-season crops of tropical or sub-tropical origins have received the most study.) All stages of growth and development of the entire plant (except perhaps the dry seed) are susceptible. This susceptibility limits the season of growth, geographic distribution, and postharvest storage conditions of these plants. (Harvested plant parts, especially fruits of some temperate plants, notably apples, pears, cranberries, asparagus, and potatoes, exhibit chilling damage during storage when exposed for extended periods at temperatures very close to freezing, i.e., around 2 °C to 3 °C. However, these temperatures do not limit growth or geographic distribution of these species as exhibited by warm-season crops.)

chilling temperature Any temperature below the *critical temperature*, but above freezing, that causes injury.

chilling tolerance The ability of *chilling-sensitive plants* or plant parts, to endure the metabolic dysfunction and/or harmful consequences that result from exposure to chilling temperatures and to survive if the abuse is not sustained beyond a certain lethal point. Chilling tolerance is used to describe this differential response to a chilling *stress* and should be confined to describing differences in the time and course of the development of *chilling injury* symptoms.

chilling treatment The process of exposure to a *chilling temperature* for a time period sufficient to cause injury.

conditioning See definition in Section 1 of this chapter.

critical temperature See definition in Section 1 of this chapter.

dysfunction An impaired functioning of plant tissues in response to *chilling*. The impairment is reversible if the tissue is returned to a nonchilling temperature after a period of exposure. However this dysfunction becomes irreversible after a longer period of time at the chilling temperature.

hardening See definition in Section 1 of this chapter.

intermittent warming Interruption of a chilling exposure with brief warm periods before the critical time is exceeded and injury occurs. Chilling-sensitive tissue can be kept for extended periods at *chilling temperatures* if the critical time is not exceeded before intermittent warming and sufficient time is spent at the warmer temperature for the tissue to recover or repair prior to returning to the chilling temperature.

primary event(s) See definition in Section 1 of this chapter.

secondary event(s) See definition in Section 1 of this chapter.

sensitivity to chilling The term used to distinguish relative sensitivity (among *chilling-sensitive plant species*) on the basis of their *critical temperature* and critical time to develop *chilling injury*.

symptoms of injury See definition in Section 1 of this chapter.

3. COLD HARDINESS[2]

anaerobic stress A stress imposed on an organism as a result of the absence of the free oxygen of air. See also *ice encasement* and *flooding*.

bacteria nucleation inhibitors Chemicals other than bactericides that inhibit ice nucleation by *ice-nucleation bacteria*.

calorie See definition in Section 1 of this chapter.

chilling requirement Low-temperature requirement to overcome dormancy in seeds and buds. (See dormancy in Chapter 17; also called **endo-dormancy**)

cold hardiness With reference to the extent that plants can survive temperatures below freezing. Can be quantitatively expressed as the *critical temperature*.

cold injury Injury incurred by biological material due to temperatures below 0 °C. The term is often used interchangeably with **winter injury**, *freezing injury* and **frost injury**.

cold protection Methods of guarding against injury from temperatures below 0 °C. The term is often used interchangeably with *frost protection* and generally refers to protection of blossoms in the spring.

cold shock Imposition of a brief, non-freezing temperature resulting in a *strain* to the organism; may or may not induce further acclimation.

convection The mass movement of heated liquid or gas. When used in discussions of cold hardiness, the term generally refers to mass movement of heated air.

critical temperature See definition in Section 1 of this chapter.

deep supercooling Ability of organisms to supercool at a temperature below that of intracellular freezing of water to as low as the homogeneous nucleation point of pure water, approximately -40 °C. This mechanism to avoid freezing exists in tissue such as xylem ray parenchyma and dormant flower buds. Refer to *supercooling*.

degree growth stage model (°GS Model) Numerical system for quantifying the annual physiological growth stages of buds of temperate plants. The annual cycle is divided into 360-degree growth stages and five major point events (0 °GS and 360 °GS = onset spring growth; 90 °GS = maturity induction point when plant first becomes responsive to photoperiod; 180 °GS = vegetative maturity and the onset of *dormancy* (see definition in Chapter 17); 270 °GS = maximum rest; 315 °GS = end of rest when chilling requirement is satisfied).

---

[2]Original author of this section was Leslie H. Fuchigami

dehardening See definition in Section 1 of this chapter.

differential thermal analysis (DTA) Method of determining the exothermic temperature difference between a reference and a sample, usually biological, being frozen, or the differential endothermic temperature of a sample being thawed.

electrical conductivity The current that will flow from one face of a unit cube of a given substance to the opposite face when a unit potential difference is maintained between these faces. A technique used to determine *electrolyte leakage* (membrane integrity) following a *stress*.

electrical impedance The application of alternating electrical current to pre-frozen or frozen plant tissues to predict damage or measure injury, respectively.

electrolyte leakage A technique used to determine cell or tissue viability by estimating membrane integrity. Following a *stress*, the tissue is shaken in a given quantity of water for a predetermined period, and the initial *electrical conductivity* of the effusate is determined. The tissue is then killed, either in liquid nitrogen or by heat, shaken for a given time, and the final conductivity determined. Viability or the extent of cellular or tissue damage is estimated by the percentage loss of electrolytes.

electrometric method (and cell freezing) Measuring the electrolyte leakage from tissue caused by membrane damage from freezing. Various electrometric methods are used; refers to *electrical conductivity*, *electrical impedance*, conductivity, and *electrolyte leakage*.

endotherms Heat consumed during a thawing event.

evaporative cooling Cooling caused by the vaporization of a liquid. In the event of cold hardiness, it would be the reduction in temperature of biological matter due to the loss of water to the atmosphere.

exotherm Heat lost during a freezing event.

extracellular freezing The crystallization of apoplastic water.

extraorgan freezing Mechanism of *freezing tolerance* in plant organs by water translocation from supercooled tissues or organs to nucleation centers in adjacent tissues (**extratissue freezing**) or outside the organs (**extraorgan freezing**).

flooding See definition in Section 4 of this chapter.

freeze avoidance Lack of water crystallization in tissues at subzero temperatures caused by the absence of either intrinsic or extrinsic nucleators. The tissue *supercools* and thereby escapes injury caused by ice formation.

freeze dehydration Loss of symplastic water caused by the vapor-pressure deficit created by ice in the extracellular spaces. The degree of dehydration is a function of temperature; as the temperature decreases, freeze-dehydration increases resulting in a concentration of the cell contents.

freeze desiccation See *freeze dehydration*.

freeze tolerance The ability of a cell or tissue to tolerate the presence of ice in the apoplastic spaces.

freezing injury Injury of biological material caused by sub-zero temperature.

freezing point The maximum temperature attained by the release of heat (*exotherm*) following the initiation of ice crystal formation. The temperature of ice crystal initiation may vary, depending on the presence of nucleating agents, but the freezing-point temperature is constant for any given solution.

freezing-point depression The lowering of the freezing-point temperature of a solution by the presence of osmotically active compounds.

frost A deposit of one of several forms of ice crystals as a result of sublimation of water vapor on the earth or earth-borne objects colder than 0 °C:

Advection Frost: Occurs from the movement of large cold air masses into an area for several days resulting in severe low temperatures and often accompanied by strong winds.

Blackfrost: A dry freeze that occurs when the dewpoint is low, preventing water-vapor crystallization on objects and resulting in the internal freezing of vegetation.

Hoarfrost or **White Frost**: A deposit of interlocking ice crystals formed by direct sublimation on objects.

Radiation Frost: Occurs on calm, clear nights when there is unimpeded radiation from the earth resulting in strong temperature inversions. Usually occurs in the early morning hours and is characterized by relatively mild subfreezing air temperatures.

frost hardening See *acclimation*. This process may be divided into three phases, based on the environmental stimulus and the type of changes occurring.

1st Stage: Triggered by short days, which stimulate the production of a *translocatable hardiness-promoting factor*; predominantly an active metabolic process.

2nd Stage: Triggered by low temperature, often a mild frost. Both metabolic and physical changes are involved.

3rd Stage: Found in hardy woody species that have been exposed to prolonged freezing temperatures, resulting in physical alterations.

frost heaving Partial or complete uplifting of a surface caused by ice expansion and resulting in exposure and injury to plants.

frost plasmolysis Contraction of the dead protoplast from the cell wall following a lethal *stress*. Primarily due to the inability of the cell to reabsorb and maintain turgor following a lethal *stress*. (See *freeze dehydration.*)

frost protection Methods of guarding against injury from temperature below 0 °C. Generally refers to protection of blossoms in the spring.

hardiness promoter(s) Naturally occurring substance(s) synthesized in plants that induce(s) freezing resistance. See also: *translocatable hardiness promoter(s)*.

heterogenous ice nucleation Nucleation that is catalyzed by a solid or liquid substrate that allows groups of adsorbed water molecules to assume configurations that are able to promote further condensation. These nucleators enhance the probability that a cluster of water molecules of critical dimensions can form, which results in crystallization.

hexagonal ($H_1$1) phase The $H_1$1 phase is basically a liquid crystal in a hydrocarbon matrix penetrated by hexagonally packed aqueous channels toward which the pole groups of the lipid are oriented. Lipids in the $H_1$1 phase provide a permeability barrier between internal and external environments.

homeohydric plant (or homoiohydric) See Section 4 of this chapter.

homogenous nucleation temperature The temperature at which a water nucleus forms that can be recognized by other water molecules as a structure resembling ice. Such clusters of molecules arise spontaneously by random fluctuations. The probability that such a cluster can serve as an effective nucleus for crystallization depends on its size and its lifetime, which are a function of temperature. Water spontaneously nucleates at -38.1 °C. The approximate homogenous nucleation temperature, T_n, for typical plant solutions is given by the equation: $\Delta T_n = -(2 \Delta T_m + 38.1\ °C)$, where ΔT_m is the melting point depression for the solution in °C.

ice deletion mutants Mutants of **ice-nucleating bacteria** that do not act as **ice nucleators.** (See *ice nucleation active bacteria.)*

ice encasement The partial or complete covering of an organism by ice resulting in anaerobic *stress.*

ice nucleation active bacteria (INA) Bacteria capable of causing ice formation at sub-zero temperatures close to 0 °C.

intercellular freezing See *extracellular freezing.*

intrinsic ice nucleator Nucleators of ice forming *within* plant tissues and organs resulting in crystallization at relatively warm subzero temperatures.

inversion layer A meteorological phenomenon in which temperature rises with increased elevation instead of falling (or decreases less than the adiabatic lapse rate would predict). Air does not rise by convection through an inversion layer.

killing temperature The temperature at which an organism cannot recover from the *stress* as measured by a viability test—often estimated with the LT_{50}.

LT_{50} The temperature at which 50 % of the population survives an artificial temperature treatment.

lamellar phase Orientation of the phospholipids with their polar-head groups exposed to an aqueous environment and the acyl side chain oriented toward the center of the phospholipid bilayer.

low temperature exotherm (*LTE*) *Exotherm* or *exotherms* that appear at a lower temperature than the large exotherm that represents freezing of extracellular water. The low temperature exotherms represent a small percentage of the total water. Low temperature exotherm (*LTE*) per se refers to the exotherm that occurs at approximately -37 °C to -40 °C, the nucleating temperature of homogeneous water.

low temperature injury Ambiguous term used for injury occurring from temperatures above freezing (**chilling injury**) to temperatures as low as -196 °C. A better term might be *freezing injury*, the temperature at which the organism no longer has *tolerance.*

melting point depression The decrease in the temperature needed to melt a solid (e.g., ice) due to the addition of solutes.

membrane fluidity A property of membranes determined by fatty-acid chain length, saturation level, and sterol components.

membrane permeability The degree to which a membrane will allow a solvent or solute to penetrate.

minimum survival temperature Temperature above which cells either tolerate or avoid freezing and below which they are killed.

phase transition temperature Lipids in membranes exist in one of several possible ordered structures in which the individual lipid molecules have more or less motional freedom. In a functional membrane, the lipids need to exist in a fluid state to allow rotational freedom and to be able to diffuse in the plane of the membrane. The temperature at which the lipids adopt a hexagonally parked structure, thus losing their diffusional freedoms, and enter a gel phase is known as the phase transition temperature.

photo oxidation See definition in Section 1 of this chapter.

plasmolysis The withdrawal of water from a plant cell causing the protoplast to contract away from the cell wall (which may shrink elastically if the cell had been under turgor).

poikilohydric plant See definition in Section 4 of this chapter.

poikilotherm An organism whose temperature varies with that of its environment.

radiation frost Heat radiated from surface of soil, trees, and any other solid particles to reduce the temperature low enough to cause frost conditions. Solid particles lose heat more rapidly than air, reducing the surface temperature. Eventually the air at low levels becomes cool and heavy compared to air at higher elevations. Occurs more readily on cloudless nights when there is no barrier for retaining the heat (i.e., a barrier that radiates the heat back to earth).

regrowth See definition in Section 1 of this chapter.

repair See definition in Section 1 of this chapter.

snow mold A disease of cereals caused by the fungus *Calonectria graminicola*. Characterized by abundant white mycelium and found beneath prolonged snow cover. A similar disease caused by the genera *Typhula*, *Sclerotium*, or *Fusarium*, which are particularly prevalent in turf-grasses.

sunscald A winter injury phenomenon in which an otherwise hardy woody plant is partially thawed on its sunward side; when the sun disappears, the thawed tissues experience rapid intracellular freezing leading to injury and death.

supercooling Also called **undercooling** or **subcooling**. Cooling of a substance (i.e., water) below the temperature at which a change of state (i.e., liquid to ice) would ordinarily take place without such a change of state occurring. Cooling of water below its freezing point without freezing taking place; this results in a **metastable state**. (See also *homogenous nucleation temperature* and *deep supercooling*.)

T_{50} or TK_{50} See LT_{50}.

thawing The process of melting of ice formed within or surrounding the organism. Ice formed within the organism may be extracellular or intracellular in nature.

theory of minimum artical cell volume Merryman proposed that cells cannot recover from a greater volume decrease than 40 % to 50 % of the unfrozen volume.

thermoavoidance An organism that escapes high temperature (e.g., by transpiration and cooling of leaves, movement of leaves in response to high temperature).

thermophile An organism that thrives at high temperature.

thermophilic Having the ability to survive or even thrive at temperatures above those considered to be lethal for most organisms.

thermophily The *tolerance* of certain organisms to temperatures between 30 °C and 100 °C.

thermostability The ability of proteins, enzymes, membranes, etc. to maintain their integrity with increasing temperature.

thermotolerance The ability of an organism or its cells or other parts (tissues and organs) to survive the full impact of a temperature stress (i.e., see *freeze tolerance*).

Tmax Temperature at which 100 % of the plants are killed.

Tmin The minimum temperature that results in injury.

translocatable hardiness promotor(s) Same as *hardiness promoter(s)*. Substance(s) produced in response to a short photoperiod that is perceived by phytochrome in leaves and translocated in the phloem to other plant parts.

viability See definition in Section 1 of this chapter.

visual browning A method of determining injury to plant tissues by observing the presence or absence of oxidative phenolic browning.

winter chilling Temperatures adequate for the **chilling requirement** to break dormancy. (See Chapter 17.)

4. WATER STRESS[3]

Despite the large body of knowledge concerning the role of water in the physiology of plants, there is a lack of unanimity among plant physiologists regarding the nature and physiological basis of plant responses to water-limited conditions. For this reason, the definitions recommended below are mainly descriptive and not intended to imply specific underlying physiological mechanisms for, or attribute adaptive value to, the terms defined. The basic approach to stress/strain follows that of Levitt (1980).

antitranspirant A substance that reduces transpirational water loss when applied to plants.

---

[3]Original author of this section was Kenneth A. Shackel.

bound water The quantity of water in a tissue that is strongly held by adsorptive (**matric**) forces and is difficult to remove by tissue drying. Since this quantity will depend on the drying conditions, the utility of a distinction between bound and free water is questionable; hence, this term is not recommended.

cavitation Spontaneous occurrence of a vapor phase in a liquid under tension. Thought to occur in the xylem of plants under moderate to severe *water stress* and to result in a reduced capacity of the xylem to conduct water.

critical period A period during crop development in which *water stress* will have a more profound effect on harvestable yield than the same stress at other periods. Can be associated with stress-related reductions in early crop growth, which will be compounded over time, or, for crops grown for their reproductive structures, direct stress-related effects on reproductive development.

dehydration (desiccation) tolerance Ability to continue a plant function (i.e., reduce *strain*) despite reductions in plant water potential (Ψ).

dehydration (desiccation) avoidance Ability to prevent reductions in plant water potential (Ψ; i.e., reduce *stress*) despite reductions in environmental water availability.

drought A condition in which the availability of soil water on a meso- or macro-environmental scale is generally insufficient to supply the maximal requirements of otherwise well-adapted plants.

drought avoidance The ability of a plant to survive or yield under drought conditions by increasing its water supply (e.g., by deep rooting) relative to other plants in the same environment. Does not imply **desiccation avoidance**.

drought escape Completion of the entire life cycle, or critical portions thereof, during drought-free periods in an otherwise drought-dominated environment.

drought resistance The ability of a plant to survive or yield under drought conditions relative to other plants in the same environment. *Drought tolerance, avoidance,* and *escape* are types of drought resistance.

drought stress The degree to which soil-water availability fails to meet the maximal requirements of otherwise well-adapted plants over a meso- or macro-environmental scale.

drought tolerance The ability of a plant to survive or yield under drought conditions despite the lack of available soil water to meet its maximal needs. Plant survival may be based on reductions in water need (e.g., stomatal closure); hence, drought tolerance does not imply **desiccation tolerance**.

evaporative demand The capacity of the aerial environment to cause water evaporation from a given object, usually a plant or plant canopy. Evaporative demand is influenced by many environmental factors including radiation load, temperature, *vapor pressure deficit* (*VPD*), and wind speed in addition to plant or plant-canopy characteristics.

flooding The partial or complete covering of an organism by water resulting in anaerobic *stress*.

hardening See definition in Section 1 of this chapter.

homoiohydric Plants that maintain a relatively constant level of hydration compared with large changes that occur in the water potential of their environment (e.g., vascular plants).

hydrophyte A plant adapted to environmental conditions of partial or full submersion in free water.

mesophyte A plant adapted to environmental conditions of moderate water availability.

osmoregulation (osmotic adjustment) Partial or complete maintenance of cell turgor over a range of total water potentials by regulated changes in cell osmotic potential. This process is well documented in the turgor and volume maintenance of certain algal species when exposed to changes in the osmotic potential of their aqueous environment. The adaptive importance of this process in vascular land plants is controversial.

paraheliotropic Orienting parallel to the incident solar radiation. Paraheliotropic leaf movements in plants under *water stress* reduces intercepted radiation and can reduce leaf temperature and leaf transpiration.

phraeatophyte A plant with the ability to thrive in a dry environment, in which high evaporative demands are met by having roots deep enough to reach a permanent water table (see *drought avoidance*).

poikilohydric Plants that equilibrate with the water potential of their environment and, for land plants, are capable of assuming a wide range of hydration states.

Relative Water Content (*RWC*) The water content of a tissue expressed as a percent of the water content of the same tissue at full hydration ($\Psi = 0$).

SPAC The concept of transpirational water flow through a **Soil-Plant-Atmosphere Continuum**, usually in terms of an Ohms-Law analogy with $\Delta\Psi$ as the potential gradient, transpiration as the flux, and the hydraulic resistance between any two points of the pathway as the ratio gradient/flux.

succulents Plants having thick leaves with a high tissue-water content. Typically *xerophytes* that have a thick cuticle, low transpiration rates, and the high *water use efficiency* (*WUE*) associated with CAM metabolism.

vapor pressure deficit (*VPD*) The difference between the actual vapor pressure of water in the air and its saturated (i.e., 100 % relative humidity) vapor pressure, expressed in pascals: Pa. It is also logical to use the term **vapor density deficit** although this is seldom done. (**Vapor density** is expressed in grams of water per cubic meter of air).

water deficit Any value of tissue Ψ that is below the highest value exhibited by that tissue in its most hydrated natural state. Does not imply *water stress*.

water stress Can refer to an environmental condition of either excess or lack of water, but the term is usually used to indicate a lack of water. Water stress as an environmental condition (sometimes referred to as *drought stress*) should be expressed as the amount of water available to a plant compared with its maximal requirements. Water stress as a physiological condition may be expressed as plant-water potential (Ψ), but the association between Ψ and water-stress-related injury is controversial. See *stress* in Section 1 of this chapter.

water status The general condition of a plant relative to any limitations imposed by a lack of water.

water use efficiency (*WUE*) A comparative measure of plant productivity per unit water used, in which the appropriate units for productivity and water use will depend on the objective of the comparison. Agricultural *WUE* can be defined for a growing season either as yield per unit irrigation water applied or biomass produced per unit of transpiration, whereas instantaneous physiological *WUE* can be defined as moles of carbon dioxide fixed by photosynthesis per mole of water transpired.

xeromorphy Morphological characteristics, such as small, tough leaves, a heavy cuticle, etc., that are generally important for adaptation to dry environmental conditions.

xerophily The combined adaptive morphological and physiological characteristics of plants able to grow under dry environmental conditions.

xerophyte A plant adapted to environmental conditions of severely restricted water availability.

5. SALINITY STRESS[4]

acclimation See definition in Section 1 of this chapter.

biosalinity Interplay of saline habitats and the organisms living within them.

cation exchange capacity (*CEC*) The total quantity of cations that a soil or other material can adsorb at a specific *p*H, usually expressed as centimoles (millimoles is preferred SI) of a specific cation per kilogram of exchanger.

chloride salinity *Salinity* in which chloride is the dominant anion.

clay dispersion The separation of clay into individual component particles resulting from the presence of monovalent cations. In saline soils, sodium is frequently responsible for the dispersion of clay. Soils with dispersed clay are dense and have a greatly reduced permeability.

compatible solute An organic solute that accumulates inside cells without causing severe metabolic disruption and may aid *osmoregulation*; for example, glycine-betaine, proline, glycerol.

cyclic salt Salt derived from the sea or salt lakes that is deposited on plants and soils from wind or rainfall.

---

[4]Original authors for this section were D.W. Rains and E.V. Maas who wish to express special thanks to Richard H. Nieman for his thorough review of and suggestions for this set of definitions.

dehydration Loss of water from plant tissues in excess of water uptake.

desalination The removal of salt from soil or water by physical or chemical means.

electrical conductivity (*EC*) A measure of the ability of a solution to conduct electrical current; it is correlated with the concentration of the ions in the solution. *EC* is used to express the concentration of salt in saline soil solutions and is given as decisiemens per meter (dS/m; millisiemens per meter is preferred SI).

euhalophyte (true halophyte) A plant whose growth is maximum at a soil water salinity of about 10 g total dissolved solids per liter ($\Psi_s \approx 0.8$ MPa; $EC \approx 20$ dS/m) and decreases at either higher or lower salinities.

exchangeable sodium percentage (*ESP*) The percentage of the total exchangeable cations that is sodium. Soils with high *ESP* can be difficult soils to reclaim.

flocculation The clumping of clay particles into aggregates as a result of the neutralization of charged surfaces of the clay.

glycophyte A plant that grows optimally in nonsaline habitats.

halomorphic soil A suborder of the intrazonal soil order, consisting of *saline* and *sodic soils* formed under imperfect drainage in arid regions and including the great soil groups **Solonchak** or Saline soils, **Solonetz** soils, and **Soloth** soils.

halophyte A plant that grows and completes its life cycle in saline habitats.

hardening See definition in Section 1 of this chapter.

ionic effect A toxic or nutritional effect of specific ions on a plant.

leaching requirement (*LR*) The amount of water in excess of the plant requirement that is needed to remove salt from the root zone in order to prevent soil salinity from exceeding a specified value.

mangroves Group of woody plants (about 11 genera) that grows in saline aquatic habitats (e.g., marine estuaries and swamps) and is capable of tolerating sea water. Common genera include *Avicennia* and *Rhizophora*.

mesophyte A plant adapted to a moderately moist habitat.

miohalophyte A plant that exhibits maximum growth in nonsaline soils but steadily decreasing growth with increasing salinity; such plants tolerate higher salinity than *glycophytes*.

oligohalophyte A plant adapted to habitats of low salinity ($\Psi_s \approx 0.04$ to 0.4 MPa at field capacity; $EC \approx 1$ to 10 dS/m).

osmoregulation Changes in cell osmotic potential that tend to maintain turgor.

osmotic adjustment Changes in concentration of certain cell solutes (see *compatible solute*) in response to changes in cell water potential. These changes contribute to *osmoregulation*. (See also *osmoregulation* in Section 4 of this chapter.)

osmotic effector See *osmoticum*.

osmotic shock *Osmotic stress* caused by a sudden and drastic change in external osmotic potential.

osmotic stress External osmotic potential below or above cell osmotic potential sufficient to cause *strain* (e.g., reduced yield). Osmotic stress almost always refers to external osmotic potentials below those in the tissues, but osmotic potentials above those of the tissue (i.e., less negative) may have some deleterious effects such as causing certain fruits (e.g., cherries) to split.

osmoticum A solute that decreases osmotic potentials of cell solutions as its concentration increases (i.e., any solute).

physiological drought Plant water deficit (i.e., low tissue water potential) caused by *salinity*, low soil water potential, or other *stress* factors.

polyol A compound containing many alcohol groups.

reclamation With reference to saline soils, the process of removing excess salt to a level that permits production of plants without significant adverse effects.

saline adaptation Genetic modification of individuals in a population that increases their ability to survive excess salt.

saline adjustment Physiological and biochemical changes of individual plants that increase their ability to survive excess salt.

saline-sodic soil A soil that is both *saline* and *sodic*.

saline soil A soil that has an electrical conductivity in a saturated-paste extract greater than 4 dS/m. The soluble salt content of such soils is sufficient to interfere with the growth of many plant species. Sensitive crop plants are affected at half this salinity and highly tolerant ones at about twice this salinity.

saline stress External salt concentrations sufficiently high to reduce plant growth. Injury may result from *osmotic stress*, ion toxicities (see *ionic effect*), and/or nutritional imbalance.

salinity Presence of inorganic ions in solution. The term often is used to refer to ion levels high enough to cause *osmotic stress*. The predominant ions include Na^+, Ca^{2+}, Mg^{2+}, Cl^-, SO_4^{2-}, and HCO_3^-. These ions may have *specific ion effects* as well as osmotic effects. Boron and other toxic substances may be present but do not contribute significantly to salinity.

salination The process of accumulation of soluble salts in the soil.

salinity threshold The maximum *salinity* a plant can tolerate without a reduction in growth below that which occurs under similar but nonsaline conditions.

salt-affected soil A soil containing excessive concentrations of soluble salts and/or *exchangeable sodium*.

salt balance A steady-state concentration of salt determined by a balance between the influx of salt and the efflux of salt from the system. This balance is frequently disturbed by inappropriate management of water and/or soils or a change in the climatic conditions.

salt glands Specialized cells or groups of cells on or near the surface of leaves or stems that secrete salt and help control levels of salt in plant cells.

salt resistance The ability of plants to take some action or employ some mechanism to avoid detrimental effects of *salinity*. The mechanism of resistance should be specified if it is known.

salt tolerance The ability of plants to cope with *salinity*.

saturated soil paste A particular mixture of soil and water. At saturation, the soil paste glistens as it reflects light, flows slightly when the container is tipped, and the paste slides freely and cleanly from a spatula.

saturation extract The solution extracted from a soil at its saturation water content.

saturation percentage The water content of a *saturated soil paste*, expressed on a dry-mass percentage basis.

sodic soil A soil that has an *exchangeable sodium percentage* (*ESP*) greater than 15. The exchangeable sodium of such soils often adversely affects soil structure and may be deleterious to plant growth.

sodicity Presence of excess sodium in soils with a *p*H range of 8.5 to 10.0.

sodium adsorption ratio (***SAR***) A relation between soluble sodium and soluble divalent cations that can be used to predict *exchangeable sodium percentage* of soil equilibrated with a given solution. It is defined as follows:

$$SAR = \frac{(Na^+)}{\left(\frac{Ca^{2+} + Mg^{2+}}{2}\right)^{\frac{1}{2}}},$$

where concentrations are expressed in millimoles per liter.

succulence 1. A morphological condition denoting thick, fleshy plant organs with a high water content. 2. Juiciness; measured as the water content per unit surface area of plant tissue. (See definition in Section 4 of this chapter.)

sulfate salinity *Salinity* in which sulfate is the major anion.

xerophyte See definition in Section 4 of this chapter.

REFERENCES

[Anonymous]. 1984. Glossary of soil science terms. Soil Sci. Soc. Amer., 677 South Segoe Road, Madison, Wisconsin. p 38.

Burke, M.J., L.V. Gusta, H.A. Quamme and C.J. Weiser. 1976. Freezing and injury in plants. Annual Review of Plant Physiology 27:507-528.

Chapman, V.J. 1960. Salt Marshes and Salt Deserts of the World. Interscience Pub., Inc. New York, p 392.

Franks, F. 1981. Biophysics and biochemistry of low temperatures and freezing. In: G.J. Morris and A. Clark, editors. Effects of Low Temperatures on Biological Membranes. Academic Press, London. p 3-19.

Levitt, J. 1980. Responses of plants to environmental stresses. Vol 1, Chilling, freezing and temperature stresses. Academic Press, New York.

Levitt, J. 1980. Responses of plants to environmental stresses, Vol 2, Water, radiation, salt, and other stresses. Academic Press, New York.

Li, P.H. and A. Sakai, editors. 1978. Plant Cold Hardiness and Freezing Stress. Vol 1. Academic Press, New York.

Li, P.H. and A. Sakai 1982. Plant Cold Hardiness and Freezing Stress. Vol 2. Academic Press, New York.

Little, R.J. and C.E. Jones. 1980. A dictionary of botany. Van Nostrand Reinhold, New York.

Luyet, B.J. 1968. The formation of ice and the physical behavior of the ice phase in aqueous solutions and in biological systems. In: J. Hawthorne and E.J. Rolfe, editors. Low Temperature Biology of Foodstuffs. Pergamon Press, New York. p 53-77.

Lyons, J.M. and R.W. Breidenbach. 1987. Chilling injury. In: J. Weichman, editor. Postharvest Physiology of Vegetables. Marcel Dekker, Inc., New York. p 305-326.

Mazur, P. 1969. Freezing injury in plants. Annual Review of Plant Physiology 20:419-448.

Raison, J.K. and J.M. Lyons. 1986. Chilling injury: a plea for uniform terminology. Plant, Cell, and Env. 9:685.

Sakai, A. and W. Larcher. 1978. Frost survival of plant responses and adaptation to freezing stress. Springer-Verlag, New York.

Salisbury, F.B., and C.W. Ross. 1992. Plant Physiology, Fourth edition. Wadsworth Publishing Co., Belmont, California.

Saltveit, M.E., Jr. and L.L. Morris. 1990. Overview on chilling injury of horticultural crops. In: C.Y. Wang, editor. Chilling Injury of Horticultural Crops. CRC Press, Inc., Boca Raton, Florida. p 3-15.

Sinclair, T.R. and M.M. Ludlow. 1985. Who taught plants thermodynamics? The unfulfilled potential of plant water potential. Aust. J. Plant Physiol. 12:213-217.

Staples, R.C. and G.H. Toenniessen, editors. 1984. Salinity Tolerance in Plants - Stategies for Crop Improvement. John Wiley & Sons, New York.

Turner, N.C. 1979. Drought resistance and adaptation to water stress in crop plants. In: Stress physiology in crop plants. H.W. Mussell and R.C. Staples, editors. Wiley (Interscience), New York.

United States Salinity Laboratory Staff. 1954. Diagnosis and Improvement of Saline and Alkali Soils. U. S. Dept. Agr. Handbook No. 60. p 160.

Weiser, C.J. 1970. Cold resistance and injury in woody plants. Science. 69:1269-1278.

Weiser, C.J. 1982. Plant Cold Hardiness and Freezing Stress. Vol 2. Academic Press, New York.

Consultants

John S. Boyer
University of Delaware
Lewes, Delaware

Kent J. Bradford
University of California
Davis, California

William J. Bramlage
University of Massachusetts
Amherst, Massachusetts

R. William Breidenbach
University of California
Davis, California

Michael J. Burke
Oregon State University
Corvallis, Oregon

Tony H.H. Chen
Oregon State University
Corvallis, Oregon

Emanuel Epstein (retired)
University of California
Davis, California

Leland E. Francois
USDA Agricultural Research Service
Riverside, California

Catherine M. Grieve
USDA Agricultural Research Service,
Riverside, California

Laurence V. Gusta
University of Saskatchewan
Saskatoon, Saskatchewan, Canada

Anthony E. Hall
University of California
Riverside, California

Robert L. Jefferies
University of Toronto
Toronto, Ontario, Canada

Delmer O. Ketchie
Washington State University
Wenatchee, Washington

Mark A. Matthews
University of California
Davis, California

Richard H. Nieman (retired)
USDA Agricultural Research Service
Riverside, California

James W. O'Leary
University of Arizona
Tucson, Arizona

Robert E. Paull
University of Hawaii at Manoa
Honolulu, Hawaii

Robert W. Pearcy
University of California
Davis, California

Albert C. Purvis
University of Georgia
Coastal Plain Experiment Station
Tifton, Georgia

Harvey A. Quamme
Research Station
Summerland, BC, Canada

James D. Rhoades
USDA Agricultural Research Service
Riverside, California

Frank B. Salisbury
Utah State University
Logan, Utah

Mikal E. Saltveit
University of California
Davis, California

Michael C. Shannon
USDA Agricultural Research Service
Riverside, California

Richard C. Staples
Cornell University
Ithaca, New York

Donald L. Suarez
USDA Agricultural Research Service
Riverside, California

Charles G. Suhayda
USDA-Agricultural Research Service
Pasadena, California

Karen Tanino
University of Saskatchewan
Saskatoon, Saskatchewan, Canada

Ralph Weimberg (retired)
USDA Agricultural Research Service
Riverside, California

Chien Yi Wang
U.S.D.A. Beltsville Agricultural Research
Center, Maryland

Conrad J. Weiser
Oregon State University
Corvallis, Oregon

Clyde Wilson
USDA Agricultural Research Service
Riverside, California

Michael Wisniewski
USDA Agricultural Research Service
Kearneysville, West Virginia

Anne F. Wrona
University of California
Holtville, California

SALISBURY
South High
1943~1993
APPENDICES
PRESENTING SCIENTIFIC DATA

APPENDICES

PRESENTING SCIENTIFIC DATA

Anyone can do experiments, but if the results are to qualify as science, more is required than special techniques, logical thought processes, and elaborate equipment. It is not the white laboratory coats and foul-smelling chemicals, or even the field stations or telescopes or orbiting satellites, that qualify a work as science. To be science, the results of one's investigations must be presented to one's colleagues for their evaluation and perhaps for them to use in their own attempts to expand the limits of human understanding. Data must be published or otherwise presented. "Publish or perish" may be thought of by some as an unfair demand placed on good teachers, but without publication (presentation), an investigation does not qualify as science. Indeed, unless the results of one's efforts are communicated to one's colleagues, the resources (often tax funds) used to produce those results are wasted.

Presentation of data requires the use of language. Typically, results are described in a scientific paper, so **Appendix A** is a discussion of writing. The discussion does not emphasize the usual format of a scientific paper (Introduction, Methods and Materials, Results, Discussion); such a pattern is well known to scientists, and specific details can be seen in the journals to which authors intend to submit their manuscripts. It is imperative that authors consult such sources before preparing and submitting manuscripts. The discussion here presents some principles of English grammar and style. Although space limitations eliminate a complete analysis of the English language, an attempt is made to cover the basics in a somewhat logical way and to emphasize a few points that often are not appreciated. Scientists all over the world must now prepare some of their manuscripts in English even when English is not their native language. Although good copy editors know the rules discussed here, many scientist authors apparently do not, and few journals do any extensive copy editing. Hence, it is easy to find examples in the current scientific literature of the problems described in Appendix A.

Data are also communicated at scientific meetings. Traditionally, this has been an oral presentation illustrated by slides. During recent years, such slide talks have included many text slides as well as the traditional photographs, data graphs, diagrams, and other figures. Computer programs have made it possible to prepare beautiful and elaborate slides. Nevertheless, such talks are sometimes poorly presented and poorly received. Members of the audience may have to struggle to understand what is being discussed. In many cases, communication could be better achieved if a few simple rules were followed during preparation and presentation of the slides. **Appendix B** discusses many of these rules. Many scientific societies now emphasize posters over oral presentations, and again, many posters are difficult to assimilate in the available time. Thus, Appendix B also discusses some rules and suggestions that can improve communication between a poster presenter and his or her audience. We recognize that writing styles, slide talks, and poster presentations represent the personal expressions of their authors. By the same token, Appendices A and B represent personal viewpoints. We hope this will not detract from their potential value.

Plant scientists often depend upon growth chambers in their research, but there are many kinds of chambers and many ways that the experimental conditions can be reported. Thus, **Appendix C** presents guidelines for measuring and reporting the environmental parameters of growth-chamber experiments. The guidelines were formulated by a special committee of the American Society of Agricultural Engineers (ASAE). The guidelines are essentially an application of the principles presented in Section II. They have been edited slightly to conform, as far as possible, to other recommendations in this book.

A

SOME SUGGESTIONS ABOUT SCIENTIFIC WRITING

Frank B. Salisbury
Plant Science Department
Utah State University
Logan, Utah 84322-4820 U.S.A.

As a budding young scientist at the university, I found my English classes to be very distasteful. There seemed to be no logic—no science—to the language, only an assemblage of arbitrary rules. Then I lived for two years in Switzerland and learned German, a highly logical language. Gradually, through analogy with the German grammar I was enjoying so much, I began to see some glimmerings of logic in my native tongue. Later, as I became involved in text-book writing, the copy editors at Wadsworth Publishing Company piqued my curiosity by the changes they made in my manuscripts. Why should *which* sometimes be changed to *that*, for example? I began to browse the style manuals and rule books, finding a little more logic in the English language, often buried in the assemblage of arbitrary rules.

For the past few years, I have been editing submitted manuscripts from the Americas and Pacific Rim countries for the Journal of Plant Physiology. Of course there are special problems faced by authors whose native tongue is not English, and part of the discussion here has them in mind. Furthermore, not all my fellow native English speakers see the structure of English as I do. Rules of punctuation, for example, based on the logic of the language as I came to understand it and as the manuals describe it, seem to be breaking down. Usage continually changes the rules.

What follows summarizes my personal view of how the logic of English can be expressed with the help of punctuation and a suitable choice of words. As far as I know, my approach to the topic is unique although I can back up every rule with references to the much more extensive manuals, and my discussion has been checked by three grammarians. I hope that you will approve of this approach and apply its recommendations. Time will tell how much usage will change the rules and hence my approach—and perhaps make the language even more arbitrary and less logical! Clearly, the best communication requires an agreement on language conventions between both reader and author. A proper comprehension of suitable writing conventions is essential, just as acceptance and understanding are required for proper use of the other units, symbols, and terms presented in this book. To this end, the following is offered.

1. THE SENTENCE

While it may be impossible to define a sentence in a broad sense that will cover all examples (Pinckert, 1986), a sentence in technical writing is seldom difficult to recognize. Indeed, grammarians recognize only six basic sentence structures (see box). A complete sentence contains at least one subject (a **noun**) with its **verb** (*Plants grow. Jesus wept.*), and most of the time it also includes an object of the verb along with modifying **adjectives, adverbs**, various phrases that act as modifiers, and often a **conjunction** that joins one word with another, one phrase with another, or one clause with another. **Prepositions** may be placed in front of nouns to show the relationship of the noun to other words in the sentence, and **pronouns** (which, in a sense, are really nouns) may be used to substitute for other nouns. Sometimes, but not often in technical writing, an **interjection** may be added although it has no real relationship to anything in the sentence. (*Ah ha, we discovered that plants grow.*) A sentence may include any of the *eight parts of speech* (written in **bold face** above), but the key to recognizing a complete sentence is to recognize the subject and the verb. The verb is needed for the **predicate** (what is being said about the subject, including the verb with or without objects, complements, or modifiers). If either the subject or the verb is missing, the result is a **sentence fragment** (or an **incomplete sentence**) rather than a complete sentence.

Writing elementary sentences seldom causes any difficulty for a scientific writer, but problems sometimes arise when a sentence contains more than one subject and/or more than one verb. Should two ideas, each with a subject and predicate, be included in one sentence, or should they be separated into two sentences? How should the relationship between these ideas be formulated and expressed? These are decisions that an author must make in the attempt to best communicate what he or she wants to say. It is essential to know the available options if one is to make the best decisions. The relationship between two ideas can be expressed at several levels, and these are indicated by various systems of punctuation.

A. Closely Related; Subject or Verb is Shared. Many authors, including technical writers, tie together (coordinate) two subject-predicate ideas with a **coordinating conjunction** and omit the subject before the second verb, knowing that the reader will refer to the original subject to understand the second predicate. The sentence just presented provides an example. The subject is *authors* (modified by several words), the first verb is *tie*, and the second verb is *omit*, which also has *authors* as its subject. The ideas are tied together by the conjunction *and*, which *coordinates* the two ideas. **Coordinating conjunctions** include *and, but, or, nor, for,* and *so*. Note that the first **clause** (a group of words with expressed or understood subject and predicate ideas) is **independent** because it has both a subject and a verb, while the second clause is **dependent** on the first. The second clause is dependent because it lacks a subject of its own but depends on the independent clause for its subject (as in this sentence, following *but*). Some dependent clauses lack a verb instead of a subject although this is less common. The question concerns how the relationship between the two ideas in such sentences should be communicated to the reader by punctuation. Since both ideas share a common subject, they are closely related, and logic suggests that no comma (no pause if spoken aloud) is needed

THE SIX SENTENCE STRUCTURES IN ENGLISH

The English sentence has six basic patterns. No matter how complicated a sentence is, it can be broken down into one or more of these designs:

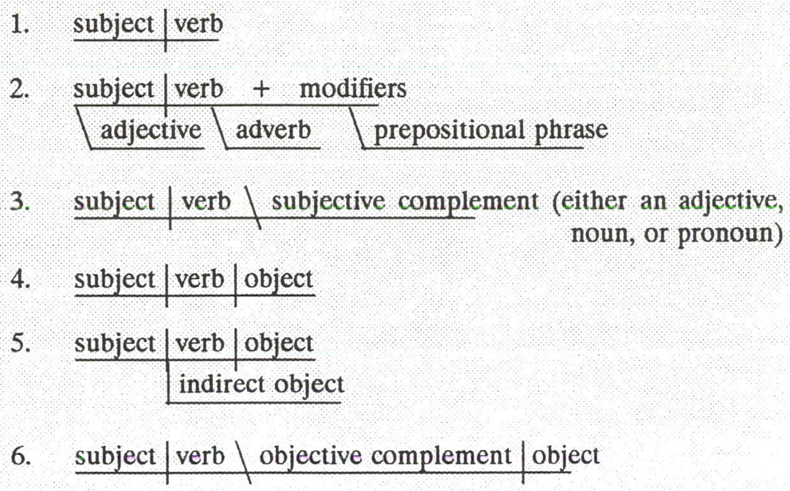

1. subject | verb
2. subject | verb + modifiers
   \ adjective \ adverb \ prepositional phrase
3. subject | verb \ subjective complement (either an adjective, noun, or pronoun)
4. subject | verb | object
5. subject | verb | object
   | indirect object
6. subject | verb \ objective complement | object

Examples:

1. *Jesus wept.*
2. *The dirty clothes are probably in the hamper.*
3. *It is cold. It tastes sour. This is he.* (The complement is either an adjective or a nominative-case noun or pronoun.)
4. *We found him. She measured the plant.* (The object is always in the objective case, a matter of concern only with pronouns.)
5. *He gave her the ring.* (The indirect object, *her*, defines the recipient of the action of a transitive verb.)
6. *The sight turned his hair grey. The speech made everyone angry. We elected him president.* (The objective complement tells what happened to the object as a result of the action of the verb.)

This material was supplied by Moyle Q. Rice.

before the conjunction. The majority of style manuals and rule books on English writing agree (although some suggest that *but* is an exception and should always be preceded by a comma).

It is common practice in modern writing, however, to insert the unneeded comma. Because *usage* rather than logic actually dictates how we should write, adding the unneeded comma may now be so common that it can hardly be called incorrect. It is *usage* that teaches a reader to understand the same conventions as the author. Nevertheless, a reader seldom if ever fails to understand such a sentence with no comma *just because* the comma is omitted. Therefore, it is logical to recommend, in spite of increasing usage to the contrary, that the comma between a dependent and an independent clause connected with a conjunction be omitted. What if the two clauses become so long and complex that the author strongly feels a need for a pause and wants to insert a comma to suggest it? Then the comma should probably be added, but it might be better to break the sentence into two sentences or to go to one of the next levels by repeating the subject and making both clauses independent; that is, avoid creating the problem.

Once the concepts are understood, punctuation should put onto paper what the author wants the reader to feel. A comma means a pause, and such a pause is usually not needed if the second clause is dependent. For example, one can read this sentence aloud and can note that no pause is needed before the *and*. The subject (*one*) of the first clause carries over to the second, dependent clause. (The two verbs are *can read* and *can note*; the second *can* could be omitted.)

If you *want* to pause for the sake of emphasis, the appropriate punctuation is a **dash** (called an em dash), which was invented just to indicate a long pause that shifts the emphasis to the end of the sentence. A dash (–) is longer than a hyphen (-) or made on a typewriter with two hyphens (--); when used like this–it is usually not set apart with spaces. It is easy to overuse the dash–and thus to weaken its effect (like that). Furthermore, dashes are seldom used in scientific writing. As a punctuation mark, the dash conveys a certain emotion, and emotion must generally be avoided in technical writing. Some grammarians suggest that the dash should *never* be used.

B. Closely Related Independent Clauses Connected by a Coordinating Conjunction. Often two more-or-less equal ideas, each expressed with both subject and verb, are tied together (coordinated) with a coordinating conjunction. In this case, both logic and the style manuals (especially in America) suggest the use of a comma before the connecting conjunction, and usage again seems to be going toward no comma. Nevertheless, the recommendation is to use the comma. I used it in the sentence about style manuals and usage, and here it is in this sentence as well. I doubt that anyone will be upset. The rule books generally say that short, independent clauses following a conjunction do not need the comma before the conjunction and that makes some sense. (Did I need a comma before that last *and*? Probably not.) Thus, there is some leeway in the use of a comma before a coordinating conjunction connecting independent clauses, but as a general rule, that comma can be helpful. It should be used more frequently than it is.

There are times when it is needed for clear understanding. Without it, a reader may have to back up, start again, and realize only on the second reading that the noun following the conjunction was the subject of a second independent clause instead of the second object of the first clause: *We measured the auxin and gibberellin was present but was not measured.* It would be well to recast that sentence, but a comma after *auxin* would help it as it stands.

C. Closely Related Sentences (no Conjunction). Sometimes an author wants to show that two sentences are closely related; they may not be related closely enough to connect them with a conjunction, however. In such a case, one has the option of simply writing two sentences, each terminated with its own **period**, or one may put a **semicolon** between them, as in the previous sentence. Of course, there are other possibilities. Would the opening sentence have been better this way: *Sometimes an author wants to show that two sentences are closely related, but they may not be related closely enough to connect them with a conjunction.* That is, *those* two ideas could have been coordinated with the conjunction *but.*

There are two kinds of errors that appear in this situation. Both should be avoided. The first is the **comma fault** (or **comma splice**) in which the two complete sentences are connected with a comma but no conjunction. This is considered a fatal error by all editors, it *must* be avoided under all circumstances. (I hope you caught it just now.)

The second error is being supported by some limited usage; indeed, there is a tendency for scientific authors to feel that they are being especially "scientific" when they use *however* instead of *but*: *An author may want to show that two sentences are closely related, however they may not be related closely enough to connect them with a conjunction.* Words like *however*, *rather*, *nevertheless*, *thus, indeed*, and others are **connectives** and not conjunctions. They are also called **conjunctive adverbs**; they both modify and connect at the same time. They are punctuated as interjections. Thus, if the author felt compelled to use *however* in that sentence, it should have been punctuated like this: *An author may want to show that two sentences are closely related; however, they may not be related closely enough to connect them with a conjunction.* That is, the second part of the compound sentence (after the semicolon) should be punctuated in the same manner as when a conjunctive adverb begins a sentence: *However, they may not be related....* In general, however, an author should realize that overuse of such connectives sounds stuffy, affected, professorial—and scientific, if that is what usage has taught us. Nevertheless, writing can often be greatly improved by eliminating such words. Surely, a reader deserves the credit for being able to see the connection between ideas without such unneeded "help." (Were connectives overused in this paragraph? Probably.)

D. Less Related Sentences. These are separated with periods. Of course they are not *un*related, which is why they are combined together in a paragraph. Separation with periods is the usual and appropriate way of writing, but it is important to have the other levels at one's fingertips in case they better communicate an author's ideas. But avoid the comma fault!

E. Two Ideas are Connected with a Subordinating Conjunction. We have been discussing situations in which two ideas are about equal to each other and are tied together in various ways (comma, semicolon) or not at all. In another situation, two clauses of *unequal* importance are tied together in the same sentence by a **subordinating conjunction**. There are many kinds of subordinating conjunctions including **relative pronouns** (*that*, *which*, *who*) and conjunctions of time (*after*, *since*, *before*, *when*, *while*, *as*, and *until*), place (*where*, *wherever*), purpose (*so that*, *in order that*), comparison (*than*, *as*, *as if*, *but if*, *as though*, *whereas*), condition (*if*, *unless*), concession (*although*, *though*, *even though*), and cause (*because*, *since*, and the weaker *for* and *as*). Any of these and others can be used to show how the subordinate clause relates to the main clause. Changing two coordinate clauses connected with a conjunction to a main clause with its subordinate clause can often clarify and generally improve the writing (Pinckert, 1986). The easy way is to use coordinate clauses. A better way, although one that requires some mental effort, is to clarify relationships by forming subordinate clauses. So that it does not sound like an afterthought, it often helps to put the subordinate clause first (as in this sentence). Consider these two sentences: *Galleys should be returned to the editor after they have been carefully read. After the galleys have been carefully read, they should be returned to the editor.*

The rule for punctuating between a main clause and its subordinate clause is simple: If the subordinate clause comes first, it is followed by a comma; if it comes after the main clause, it is not. (*...it is not if it comes after the main clause.*) Because a subordinate clause should never stand alone (forming a **sentence fragment** rather than a complete sentence), it must always occur in the same sentence as its main clause. This means that it should not be separated from a preceding main clause by a comma or semicolon. That is the rule, although it sometimes seems appropriate to add the comma for emphasis (as was done here). In the most informal writing, an author can add a dash to provide strong emphasis on the subordinate phrase or clause—although dashes tend to be overused by writers who are insecure in their knowledge of punctuation. When in doubt, add a dash! Of course this should be avoided.

There is a complication. Some subordinating conjunctions (e.g., *although*) can be used as conjunctive adverbs. In such cases, punctuation should follow the rules described above for such conjunctive adverbs.

F. Beginning a Sentence with a Coordinating Conjunction. At least one important question remains: Is it correct to begin a sentence with a coordinating conjunction (*and*, *but*, etc.)? Because coordinating conjunctions are normally used to splice together two clauses, they cannot come at the beginning of the sentence if both clauses are present (as the *sub*ordinating conjunction *because* did in this sentence). But what about a sentence that begins with a conjunction but contains only one clause? Sometimes, for emphasis, an author may want to arbitrarily make two sentences out of a compound sentence that has a coordinating conjunction between two independent clauses. Doing so ties the sentence that begins with the coordinating conjunction to the thought of the previous sentence more closely than would be the case without the conjunction. *It is acceptable to begin a sentence with a conjunction. And sometimes it can provide impact.* In this case, the coordinating conjunction acts more like a conjunctive adverb. But like many devices of this type, it can be overdone. It should be used with care.

2. MODIFYING WORDS

For the most part, the use of modifying words is relatively easy in the English language, but a few problems arise.

A. Adjectives. Adjectives are words that modify (explain) nouns and pronouns. Because they are not declined according to gender and case as in many other languages, their use in English is quite simple. The few problems that appear in technical writing concern use of the hyphen in compound adjectives, use of the comma when more than one adjective modifies a noun, and distinguishing between adjectives and adverbs. (Some languages don't use **articles**, which are special adjectives that limit or give definiteness to the things represented by the noun. *A* and *an* are indefinite articles; *the* is the definite article: *A beaker* says there is only one beaker but doesn't specify which one; *the beaker* and *the beakers* refer to a specific beaker or beakers that may have already been mentioned.)

A **compound adjective** consists of two or more words that act as a single idea to modify a noun. The compound adjective is usually formed by placing a hyphen between the two or more words (*a four-year-old tree*).

The two words might both be adjectives, one might be an adjective and one a noun, or both might be nouns, but the combination acts as an adjective: *a cold-water shower*. Even adverbs can be part of compound adjectives. To qualify as a compound adjective, the words must act together in a special way rather than each modifying the noun independently. *Far-red light* provides a good example. We are not talking about a distant, red light; instead, we are discussing light in the *far-red* part of the spectrum. Usually, when two or more adjectives precede a noun, each modifies the noun independently: *a bright, red, metal-halide lamp*. These are called **coordinate adjectives**, and in the example, one is a compound adjective that consists of two nouns (*metal* and *halide*) acting together as an adjective that describes the lamp.

It is sometimes difficult to know whether a comma should be used between a series of coordinate adjectives. Current usage often tends to eliminate commas, but in technical writing it helps to use them. If there are more than two coordinate adjectives, commas become mandatory; if there are only two, a helpful device is to see whether the insertion of *and* is acceptable. If it is, the two adjectives are quite independent of each other, and the comma should be added. If the *and* sounds out of place, the two adjectives are related to each other *almost* as they would be in a compound adjective; the comma should be omitted. If *a bright and green light* doesn't sound right, leave out the comma: *a bright green light*. If *a hot and red light* is acceptable, use the comma: *a hot, red light*. Clearly, the distinctions can be fine, and decisions must be left to the author, but the author must be aware of the accepted conventions.

Two words in a compound modifier might be an adjective and a noun, with the adjective and the noun acting together to modify another noun: a *short-day plant*. This does not say that the plant is short or that it is a day plant but that the plant responds to *short days*. Note that the combination *short days* by itself does not form a compound adjective but is simply a case of *short* acting as an ordinary adjective to modify the word *day*. The hyphen is appropriate only when the combination acts together as an adjective to modify another noun (which is sometimes implied but not stated: *irradiate the seeds with far-red*).

It is appropriate in scientific writing to combine a number (an adjective) with a unit (a noun) to form a compound modifier *when the number and unit are written out: a four-hour night, a ten-milliliter aliquot, a five-gram sample, etc.* (But, note: *four hours, ten milliliters, five grams, etc.*) As discussed in Chapter 1, however, when numerals and unit symbols are used, the hyphen should be omitted. This is because the unit symbol should be thought of as a mathematical symbol that stands for a physical quantity. Thus, a 0.5 kg sample is the mass of the standard kilogram in Sevres, France, multiplied by 0.5. Use of the hyphen destroys this relationship. When numerals and symbols are spelled out, the rules of grammar apply; when their symbols are used, mathematical rules apply.

Sometimes two or even three nouns act together as an adjective. Logic would suggest that they should be connected with a hyphen, but usage often makes the hyphen seem out of place: *cell-wall materials* or *cell wall materials*? Most authors would use the second form, and most editors would probably change the first to the second form. But stand up for logic if you feel so inclined, even if your creation is likely to be changed by an editor.

Adjectives can be the present or past participles of verbs, in which form they often appear in compound adjectives, especially in scientific writing: *a light-controlled switch* (**past participle**); *a light-controlling switch* (**present participle**). Note that the meanings are different in these two cases. In the first, light controls the switch; in the second, the switch controls the light. Although these forms are common in technical writing, the comparable longer constructions are often easier to understand: *a switch that light controls; a switch that controls the light.*

To summarize some recommendations for the use of hyphens in compound modifiers: Do not use with a combination of numbers and units that modifies a noun (*a 100 W lamp*). Use with nearly all other compound modifiers; this will make it easier for most readers and offend only those who have decided that punctuation should be eliminated just for the sake of eliminating punctuation. Avoid adding the hyphen between some compound modifiers that are often seen without it (*the high school reunion*), usually consisting of two nouns rather than an adjective and a noun. Do not use a hyphen between an adverb and an adjective if the adverb ends in *-ly*. Otherwise, compound adjectives that include adverbs are often hyphenated: *a well-done steak* (but *a steak that was well done*), *a well-known expert, a loose-fitting garment* (but *a loosely fitting coat*), *a well-guarded secret* (but *a carefully guarded secret*). All compound numbers are hyphenated (*sixty-seven samples* and *there were only sixty-seven*), but chemical terms never are: *hydrogen sulfide gas.*

Some words have prefixes or suffixes that are always hyphenated: *self-rule, ex-husband.* Others are combined with the word and never hyphenated: *overdose, underground, coworker.* When in doubt, consult a dictionary.

B. Adverbs. An adverb can modify verbs, adverbs, and adjectives; it is a versatile part of speech. If a modifier is modifying anything but a noun or pronoun, it is an adverb. Adverbs tell where, when, how, or how much. Many adverbs end in *-ly*, but many do not, and not all words ending in *-ly* are adverbs. Minor trouble arises when adverbs and adjectives are confused, but this occurs most often in speech and not in technical writing. We naturally write: *The good experiment was done well.* Not: *The* *well experiment was done good. Well* is the adverb that modifies the verb *was done*, while *good* is the adjective that modifies the noun *experiment*.

Note that **sense-impression verbs** are followed by the adjectival rather than adverbial forms: *It tastes good. It smells bad. He looks sick. I feel bad* (rather than *I feel badly*). *I feel good.* (*I feel well* means that I feel *healthy* rather than sick if *well* modifies *I*–or that I have a good sense of touch if *well* modifies *feel*.)

A word of warning about adverbs: Many writers seem to have the notion that adverbs are very elegant, but often they are inherently vague. How elegant is *very* elegant? How vague is *inherently* vague? We can often tighten our writing by eliminating *very*, *too*, *greatly*, *really*, *actually*, *extremely*, *quite*, *rather*, *slightly*, *fairly*, *somewhat*, *to a certain extent*, and very many others! Of course these words are sometimes useful.

C. Prepositions and the Objective Case. English is relatively easy for foreigners to learn (to begin with, at least) because nouns, as well as adjectives, are not inflected according to case and gender. This means that there are few problems with the use of **prepositions**, which are words that are placed in front of nouns and pronouns to show the relationship of the noun to other words in the sentence: fill *to* the calibration mark, *for* a good reason, *behind* the scutellum, *with* care, *between* the lines, etc. Nouns following prepositions are always in the **objective case**, but since English nouns have the same form in the subjective and the objective cases, we seldom give the matter any thought. That is probably why we may have trouble when we use English pronouns that do have a special form in the objective case. There are only six of them in modern English: *me*, *him*, *her*, *us*, *them*, and *whom* (plus *whomever* and the obsolete *thee*).[1]

We seldom have trouble when the noun follows the preposition directly, but many Americans (who seem to have little sense of case because it is such a small part of the language) say *between you and I* instead of the correct *between you and me*, and it is not uncommon to hear *with we girls* instead of *with us girls*. Direct or indirect objects of verbs are also in the objective case: *He gave her the unit, and its probe contacted him and me.* In technical writing, we use fewer personal pronouns anyway, but we must be careful when we do use them.

D. Personal Pronouns. As a matter of fact, we should probably use more personal pronouns in our scientific writing. It is highly artificial to put our writing always in the third person by saying *the author* did this or that instead of simply saying *I* or *we* did it. Many modern editors are now insisting on the first person instead of the out-of-date third person. For one thing, an author should be willing to take responsibility for his or her experimental results and philosophical suggestions by speaking in the first person instead of hiding behind some almost anonymous *author* who seems to be doing the writing. Saying *the author* doesn't make a paper any more objective.

---

[1]The subjective forms are *I*, *he*, *she*, *we*, *they*, *who*, and *thou*; possessive forms are *my, mine*, *his*, *her, hers, its, our, ours*, *their, theirs*, *whose*, and *thine*; *it* and *you* are the same in the subjective and the objective cases.

3. MODIFYING PHRASES AND CLAUSES

A. Restrictive and Nonrestrictive Phrases and Clauses. Problems encountered with phrases are mostly concerned with how they should be punctuated. Often the author must decide how to punctuate, depending on the exact meaning that he or she wants to convey. The key to making the decision correctly is to understand the concept of the restrictive versus the nonrestrictive phrase or clause. A **restrictive** phrase or clause contains information that is essential to understanding some part of the sentence not included in the phrase or clause; a **nonrestrictive** phrase or clause adds information that may well be important but that is not essential to understand the parts of the sentence not included in the phrase or clause.

These two possibilities can often be illustrated with the same sentence, which is why the author must decide which one expresses his or her intention. *My son, the doctor, sent me a letter. The doctor* is a nonrestrictive phrase that suggests that the speaker has only one son, who happens to be a doctor. *My son the doctor sent me a letter.* In this case, the speaker has more than one son, but the restrictive phrase *the doctor* tells which son sent the letter. The difference is the punctuation. Nonrestrictive words, phrases, or clauses are set off by commas; restrictive ones are not. Here is another example: *The girl standing in the garden waved to him.* You can identify the girl only by knowing that she was standing in the garden. *The girl, standing in the garden, waved to him.* It happens that the girl who waved was standing in the garden, a bit of interesting but nonessential information.

In correct application, the use of the **relative pronouns** *which* or *that* depends upon whether the clause that is introduced is restrictive or nonrestrictive. *Which* is the nonrestrictive relative pronoun; *that* is restrictive. *The automobile that was speeding was completely destroyed.* To know which automobile was destroyed, we must know that it was speeding. *The automobile, which was speeding, was completely destroyed.* The important thing that the sentence tells us is that the automobile–no question which one–was destroyed; as it happens, it was speeding. In our reading, we gain the sense of whether a phrase or clause is nonrestrictive or restrictive by the use or omission of the comma or commas. When there are relative pronouns, the use of *which* or *that* fortifies that sense of restrictiveness or nonrestrictiveness.

In scientific writing, most phrases or clauses introduced by relative pronouns are restrictive; such phrases should be introduced by *that* without a preceding comma. When an occasional nonrestrictive phrase or clause is used, it should be introduced by *which*, proceeded by a comma. To improve one's writing, an author should engage in a "which hunt," replacing restrictive *whichs* with *thats*.

This seems like a simple enough rule to follow, and in scientific writing the concept of restrictive or nonrestrictive can often convey important information that an author might want the reader to comprehend. Judging by published technical articles, however, many authors and editors are unaware of the rule. It is only consistently followed in the best-edited, non-technical magazines or books. Apparently many writers have the mistaken idea that *which* is more formal than *that* and thus should be used in the most formal or technical writing. The use of these two relative pronouns has little or nothing to do with formality; it is all a matter of restrictiveness or nonrestrictiveness. The recommendation is to follow the rule but not to get too upset when others fail to follow it. Perhaps, as the years go by, it will be adhered to by more and more authors and editors.

That and *who* provide another pair of relative pronouns. Again, the rule is simple, and again it is often broken. *Who* should be used with reference to people; *that* is used for everything else. A clause beginning with *who* (or *whom* if it is the object of a sentence or a preposition) can be restrictive or nonrestrictive; use of the comma tells when it is nonrestrictive.

Nonrestrictive **introductory phrases** or phrases at the end of a sentence should be set off from the rest of the sentence with a comma in formal writing. Usage has led to the omission of this comma when the phrases are short, but more frequent use of a comma to set off introductory or final phrases would lead to clearer writing. As reading experience demonstrates, the slight break or pause indicated by the comma often contributes to ease of understanding, as in this sentence.

The discussion of how to punctuate nonrestrictive phrases or clauses brings up a logical rule that is often violated by modern writers: **Put a pair of commas, or none, between subject and verb, or verb and object, or subjective complement.** To say it another way: A subject and its verb should never be separated by a single comma (unless the comma occurs between coordinate adjectives). If subject and verb are separated by a nonrestrictive phrase or clause, there must be two commas that surround the phrase or clause. The rule is broken, especially in technical writing, because the subject may be modified with so many words and phrases that the author feels the reader will run out of breath by the time he or she gets to the verb; the rest afforded by a comma seems to be in order. But this comma will be very distracting to a reader who is really paying attention. The reader is anticipating the action, the verb, and is confused by being told to pause just before getting there when no nonrestrictive (parenthetical) material justifies the pause. *The **sample** that had a large, green leaf attached to the brown stem with an expanded petiole base but virtually no thickened cuticle or acute lobe **was** chosen for the herbarium.* Some authors might be tempted to put a comma before the *was*. Of course, the sentence would be improved by recasting it as two sentences.

B. Parenthetical Phrases or Clauses. Nonrestrictive phrases or clauses (as just described) are **parenthetical**, which is to say that they contribute important information but are not essential to understand the rest of the sentence or to its grammatical structure. If the sentence is correctly constructed, the parenthetical phrase or clause can be removed (along with the punctuation that sets it apart, the parentheses in this case), and what remains will still be a grammatically correct sentence. Parenthetical phrases can be punctuated in four ways: with commas (as we have been discussing), with parentheses[2] (round brackets), with brackets (square brackets), and with dashes. The choice belongs to the author, but the choice can convey an author's feeling about the parenthetical material. If commas are used, the information is closely related to the sentence, almost restrictive, we might say.

---

[2]In Britain and other United Kingdom countries, the term *parentheses* is generally applied simply to portions of text that are parenthetical while *bracket* is a generic term for all parenthetical symbols: round brackets( ), square brackets[ ], curly brackets{ }, and angled brackets< >.

If **parentheses (round brackets)** are used, the material is more of a side issue. It is important (or it would not be included at all), but it is not as closely related to the rest of the sentence as would be the case when commas are used. Used too often, parentheses can be distracting, always confronting the reader with extraneous information that may seem beside the point. This feeling can often be changed just by changing some parentheses to commas. If the information really is almost unrelated to the rest of the sentence, however, it should be in parentheses, and a long sentence can sometimes be made easier to understand by placing some of the material in parentheses. In a sense, this removes it from the sentence although it might remain exactly where it was. (The material might also be moved somewhere else.)

Brackets [square brackets] are often reserved in formal writing for comments inserted by an editor. The editor can be the author if he or she is quoting someone else but needs to insert an explanation or comment in the quoted material. Brackets can also be used as parentheses within parentheses: *Evans and an assistant (Gillespie, who made her own study of a flightless bird [the kiwi] in Australia) spent several difficult months in the field.*

Dashes are much less formal and should seldom if ever be used in technical writing–unless the author feels justified in adding the strong emphasis provided by use of the dash or dashes–and if the author is confident that the editor will not remove the dashes! If commas are included in a phrase set off with dashes, the dashes become essential. (The exclamation point is also seldom used in technical writing–for the same reasons dashes are seldom used!! [I tend to overuse both!])

Two points about parenthetical material need to be noted: First, since all parenthetical material is by definition nonrestrictive, parenthetical phrases (regardless of how they are punctuated) that are introduced by a relative pronoun should always use *which* (or *who* or *whom*) instead of *that*. Second, since it should always be possible to remove a parenthetical phrase or clause without affecting the structure of what is left, an author must never use double commas, (like this), around parentheses or brackets.

Where is the period placed in relation to material in parentheses or brackets? If the material in parentheses comes at the end of a sentence and is itself an incomplete sentence (sentence fragment), the period is placed outside of the parentheses (like this). If the material in the parentheses comes at the end of a sentence but by itself forms a complete sentence, then such a parenthetical sentence should be set within its own parenthesis. (In such a case, the first letter of the parenthetical sentence should be capitalized, and a period should be placed at the end of the sentence and before the last parenthesis, like this.) There is no rule saying that parenthetical material must be included in some other sentence; it can and often should stand on its own, as in the example. There is also no rule saying that parenthetical material cannot form a complete sentence within a sentence (it can be distracting, as here, so it is well to avoid the practice when possible), but if the complete-sentence, parenthetical material can be placed after its "parent sentence," it might just as well be given a life of its own, cut off from its parent. When it is included within another sentence, it is not punctuated as an independent sentence.

4. VERBS

A. Plural and singular verbs. Plural verbs must be used with plural subjects, singular verbs with singular subjects. That is, a verb must agree with its subject in person and number.

A subject consisting of two or more singular nouns connected with *and* is plural: One noun and another noun make a plural subject. If two singular nouns are connected by *or*, the subject is singular (but plural if the noun closest to the verb is plural).

A singular *subject* followed by a modifying prepositional or other phrase that contains plural nouns or more than one singular noun *is* nevertheless singular (as in this sentence and the one that begins the previous paragraph).

Some nouns taken from languages other than English form their plurals in ways that are not always familiar; watch for these (*datum* and *data, medium* and *media,* etc.; see Section 8).

B. Verb tense. Verb tense should be consistent. It is usually logical to use the past tense in describing methods, materials, and results in a scientific paper: *We **found** that applied IAA strongly **promoted** elongation of intact pea plants.* The experiments were done in the past, and it is conceivable that they might give different results if repeated (if *all* determining conditions are not known or understood). Hence, the honest way to describe them is to use the past tense. Avoid changing tense in the middle of a description of methods or results, usually in a single paragraph. Published results may be described with the present tense: *Yang et al. (1993) showed that a continuous supply of auxin **enhances** stem elongation in intact plants.*

C. Participles. English forms a **present participle** by adding *-ing* to the infinitive of the verb. This is combined with some form of the verb *to be* to emphasize an action that *is occurring* (or that *was occurring* or *has been occurring* or *will be occurring*). This is such an important part of the English language that native speakers virtually never use it incorrectly, but it is often difficult for writers whose native tongue is not English. The tendency is to use this verb form too often, when it is not needed. Even native speakers can frequently tighten their writing by changing to the simple forms (*it occurs*, *occurred*, *has occurred*, *will occur*, for example).

A special problem is the **dangling participle**, which is a participle that cannot be connected immediately and unmistakably with the word(s) to which it refers. Because the **antecedent** of the verb is often left to the reader's imagination, sentences with dangling participles can often be quite ludicrous: *Coming into the greenhouse, the large skunk cabbage gave off an overwhelming stench.* (Who entered the greenhouse? The writer or the skunk cabbage?)

English usually forms a **past participle** by adding *-ed* to the verb and combining it with another auxiliary verb, usually a form of *to have.* This verb form indicates that an action was begun in the past relative to the time being referred to but is completed in that time being referred to, which can be the present, the past, or even the future: *I have measured. She has measured. He **had** measured. You **will have** measured.* The past participle can usually be replaced with a simple past or future form: *I measured. She measured. He measured. You will measure.* Nevertheless, it is sometimes appropriate to use the past-participle form. This form usually causes no trouble for native speakers unless the past participle is irregular or differs from the simple past, also formed by adding *-ed.* Examples of irregular forms include: *I give it. I gave it. I have given it. I do it. I did it. I have done it. She comes home. She came home. She has come home. He goes. He went. He has gone.* Consult a dictionary if in doubt about the past participle.

D. Passive voice. Authors of journal articles often use the **passive voice** by combining the past participle of a transitive verb with any form of the verb *to be: The water potential* ***was measured.*** This construction omits the perpetrator of the action, the one who did the measuring in the example. The first-person personal pronouns are avoided so the writing may seem more objective or "scientific." (Did you catch the passive voice in that sentence?) Actually, use of the passive voice allows one to avoid responsibility for one's actions, to divorce the worker from his or her work. Try to use the **active voice**: Instead of saying, *the leaves were treated* (passive voice), say *we treated the leaves* (active voice).

5. SOME FURTHER NOTES ABOUT PUNCTUATION

The discussion so far has followed my "logical" approach to the structure of the English language with its simple-to-compound sentences, restrictive and nonrestrictive modifiers (words and phrases), and special verb forms. This discussion has placed many items of punctuation into that "logical" context, but several important although often arbitrary rules did not fit into that evolving but somewhat limited discussion. Thus the following brief outline (based on Pinckert, 1986) summarizes the important rules of punctuation; items already discussed in some detail are presented in small type:

A. Avoid sentence fragments (incomplete sentences terminated with a **period**), and do not use a **comma** when a period (or **semicolon**) should be used; that is, avoid the comma fault.

B. End questions with a **question mark,** but do not use a question mark with indirectly quoted questions: *She asked who Bob was.*

C. Reserve the **exclamation point** for true exclamations or commands (*Drop dead!*), which are seldom used in technical writing! When the exclamation point is used for emphasis, a reader soon gets tired of such exclamatory writing! Its use suggests sentence failure! Avoid using the exclamation point just for emphasis! (But its use in the last sentence is valid because that is a direct command, an **imperative**.)

D. Use the **semicolon:**

- between independent clauses not connected with a coordinating conjunction.
- to separate complex items in series when each item itself consists of a series of items: *These are important plant hormones: auxin, typically a stem growth promoter; the gibberellins, also promoters of stem growth; cytokinins, stimulators of cell division; and ethylene and abscisic acid, sometimes called stress hormones.*

E. Use **commas**:

- between items in a series, including before the *and* that precedes the last item: *The equipment included a camera, portrait lens, and flash attachment.* This **serial comma**, as it is called, is often omitted by modern writers. Most manuals still recommend its use.

- after many adverbs, phrases, and clauses that introduce a sentence and are followed by a voice pause: *To avoid precipitation, it is essential to stir continually.*
- to enclose nonrestrictive phrases and clauses.
- before a coordinating conjunction introducing an independent clause.
- to set off direct address and other parenthetical interrupters. *"Carol, will you fix breakfast?"*
- to introduce or interrupt short quotations. *"Not," she said, "on your life."*
- to prevent misreading. *"Let's talk about Prof. Jones, and good scientists."* Such **style commas** can be inserted to emphasize an adverb (*She snored, heavily.*) or to give extra emphasis to modifiers of equal importance.

F. Use the **dash** to indicate an extended pause and give emphasis—but seldom if at all in technical writing.

G. Use **commas, parentheses, dashes**, and **brackets** to set off parenthetical material.

H. Use **ellipses** (a series of three periods) to show that something has been omitted from a quotation. If the ellipses come at the end of the sentence, add the fourth period. Use **brackets** [square brackets] to show that something was inserted by an editor.

I. When used as a mark of punctuation within a sentence, a **colon** means *as follows, an example follows, here is the explanation or list*, or *here is what he or she said*. Whatever follows a colon refers back to what immediately preceded it. It is logical to use a capital letter after a colon if a complete sentence follows, but many editors will change the capital letter to lower case. If a list follows a colon, the words are not capitalized unless they are proper nouns or must otherwise be capitalized. The colon is also used in several other ways in scientific writing as often noted in this book (e.g., to show the proportions of components in a mixture used in chromatography: water:acetic acid:butanol 5:1:3).

J. The **apostrophe** is used to form contractions (*don't worry*) and possessives (*Joule's book*; originally a contraction of *Joule his book*). If the possessive is a plural ending in *s* or a *z* sound, the apostrophe comes after the *s*; some manuals suggest adding another *s*, but because there is so much disagreement, this can be a matter of personal choice. One suggestion is that the *s* should be added if it is easily pronounced: *Descartes's essays* (because neither *s* is pronounced in *Descartes*). Usually the extra *s* is not added in technical writing. Indeed, contractions are used less often in technical writing than in less formal writing.

K. Quotation marks are used around direct quotations but not around indirect quotations. If the quotation extends over more than one paragraph, quotation marks begin each paragraph but end only the last paragraph. Quotation marks are used to set apart a word or phrase that is used in some sense other than the usually accepted one, but this is easily overdone and should be avoided as much as possible. Put commas and periods inside closing quotation marks; put semicolons and colons outside. Other punctuation (questions marks and exclamation points) should be put inside the closing quotation marks only when the punctuation is actually part of the matter being quoted: *He asked: "Should the catalyst be added?" Is it really, as written, "larger than the primary leaf"?*

L. Hyphens are used:

- in compound adjectives.

- to form compound nouns, although it is often acceptable to combine the two nouns into one word without the hyphen: *mailman* instead of *mail-man*, for example.
- to divide words at the end of the line, but rules for hyphenation are too involved for summary here. When in doubt, check in a good dictionary–or trust your word processor's hyphenation feature.

M. The **slash** is used in several technical applications, as noted elsewhere in this booklet (e.g., in fractions: 1/20). It is also sometimes used in the expression *...and/or...* in scientific writing. Some editors do not like this and will suggest *...or...or both*.

N. Underlining is used in place of ***italics*** in type-written or hand-written material. There are instances in scientific writing where italics or underlining must be used, as in scientific names of organisms (see Chapter 2). Italic type has been used appropriately in this appendix to set off *examples*. Italics can also be used for emphasis, but this is discouraged in technical writing. Foreign words are underlined or italicized, especially if they are unfamiliar.

6. ABBREVIATIONS

Science speaks its own language with specialized words that we must all learn in our respective fields. Perhaps that is why we always seem to want to invent even more terms by constructing abbreviations. Of course many abbreviations or acronyms are recognized; they are already a part of the language of plant physiology: ATP, DNA, IAA, 2,4-D, NADP, SDP, and many more. Journals often publish lists of such accepted abbreviations and expect authors to use them. (Note tables in Chapter 10.) But it is an imposition to expect a reader to learn a handful of new abbreviations in order to read a paper. A few abbreviations (four or five?) may be justified to avoid the constant repetition of some complex terms or phrases, but most authors will achieve a more sympathetic audience if new abbreviations are kept to a bare minimum. It is also helpful if the newly introduced abbreviations are easily distinguished from each other. Consider, for example, a series of treatments with or without auxin, light or dark, at morning, noon, or night: *ALM, ALN, ALNi, NALN, NALN, NALNi, ADM, ADN, ADNi, NADM, NADN,* and *NADNi*. Pitty the reader!

7. UNNECESSARY WORDS

We have a great tendency to expand our writing by using words that are not needed or that have shorter and more concise counterparts (Heichel et. al., 1990).

1. Some words can simply be dropped (e.g., *simply* in this case):
   prior history (all history is prior)
   careful study, *careful* examination (how else would you do it?)
   very (this word only contributes something in certain negative constructions:
   It isn't very effective.)
   it is shown that (seldom needed)
   it is a fact that (seldom needed)
   it is emphasized that (seldom needed)
   it is known that (seldom needed)

2. Some words can be replaced by more descriptive and concise terms:

Instead of:	**Use:**
in the absence of	without
higher in comparison to	more than
was found to be	was
in the event that	if
small number of	few
was variable	varied
additional	added, more, or other
approximately	about
at the present time	now
at that point in time	at that time
establish	show
identify	find, name, or show
in a timely manner	promptly
necessitate	cause or need
appears to be	seems (also overused, often to avoid making a factual statement: *It seems to be raining.* Better: *It is raining.*)

8. WORDS WITH SPECIAL PROBLEMS

Some words are used incorrectly so often in scientific writing that some readers may never learn to use them correctly when they become authors. Some of these are **homonyms**, which are pairs or groups of words with identical or similar sounds but different spellings and meanings. The following list includes some of the problem words including a few homonyms. Hopefully, the list includes many of the words that pose special problems for speakers whose native tongue is not English. If you are in doubt about the correct meaning and spelling of a word, check with a good dictionary or in a manual that lists such words (e.g., CBE Style Manual Committee, 1994; de Mello Vianna, 1977; Strunk and White, 1979).

accept, except *Accept* means to receive or admit, to regard as right or true, or to bear up under. *Except* means to leave out, exclude, or excuse.

affect, effect Each can be used either as a verb or as a noun, but most of the time, *affect* is used as a verb (to influence or cause a change in) and effect is used as a noun (a result or consequence of an action). As a noun, *affect* is used only as a technical term in psychology. *Effect* is used as a verb in the sense of *to cause* (*to bring about or make*). Technical writers like to sound scientific by using *effect* as a verb, but it often leads to confusion, and it is easy to say *cause* instead. *Effect* as a verb sounds affected. Furthermore, *affect* is a rather weak verb. It is better to tell *how* something is affected: *increased* or *decreased*, for example.

as When *as* is a preposition, meaning in the role, capacity, or function of (see *like, as* below), it is always followed by a noun or pronoun in the objective case. Otherwise, the case can depend on the sense that is desired: *You need her as much as I* (i.e., *as much as I need her*). *You need her as much as me* (i.e., *as much as you need me*). Ambiguity can result when *as* is used as a conjunction instead of *since* or *because*: *She did not hear the bell as she was on the terrace.* Did she fail to hear the bell *because* she was on the terrace or *while* she was on the terrace?

as...as, so...as In positive comparisons, *as...as* is the construction that is used: *as tough as nails.* In negative comparisons, either *as...as* or *so...as* can be used: *...not as* (or *so*) *skilled as his technician.* The first *as* should not be omitted in positive comparisons. (Don't say: *The sound was clear as a bell.* Say: *...as clear as a bell.*)

can, may *Can* is used to indicate ability to do something; *may*, to ask, grant, or deny permission to do it. This distinction should be followed in formal writing.

datum, data Traditionally, especially in technical writing, *datum* has been considered singular (a fact or single item of information; a single number) and *data* plural, but popular usage has almost eliminated the singular *datum* from the language, and *data* is almost universally used as a singular noun (a collective: information organized for analysis). A few of us continue to say *...these data are...*, but as our generation dies off, *data* will no doubt be used only as a singular noun. (I find this regrettable!)

due to This expression is often overused in technical writing. It is correct when it is used as a predicate adjective that follows some form of the verb *to be* and in the sense of *caused by* or *attributable to*: *The broken centrifuge was largely due to faulty maintenance.* This could be replaced by *...was caused largely by...* It is somewhat less correct, although commonly used, in the sense of *because of*, *on account of*, *owing to*, or *through*: *The centrifuge failed due to faulty maintenance.* In formal writing, it would be better to say *...because of faulty...*

its, it's *Its* is the possesive form of the pronoun *it*, but used in this case without the apostrophe: *...the graph with its curves...* *It's* is a contraction of *it is* and *it has: It's not new; it's been done before.* (Overuse of the contractions sometimes makes the writing seem too informal.)

et al. This is properly used in bibliographies to mean *and others*. Note the period after the second element. Because it is Latin, some editors insist that it be *italicized* (or underlined).

information In some languages other than English, the comparable term for *information* may be a plural (e.g., French, Spanish, Russian) or may be used as a plural (e.g., German). It is *never* correct in scientific English to use *information* as a plural: *informations.*

lay, lie *Lay* (to put, place, or prepare) always takes a direct object (*lay it down*); that is, *lay* is a **transitive verb**. *Lie* (to recline or be situated) never does; that is, *lie* is an **intransitive verb**. But the past tense and past participle of *lay* is *laid*, and the past tense of *lie* is *lay*; the past participle is *lain*. This certainly leads to confusion. *Sit* and *set* are equally troublesome.

like, as *Like* and *as* are correctly used as prepositions expressing different senses. In this case, *like* indicates resemblance to the object mentioned: *He looks like his brother.* It can always be replaced by *similar to*. *As* indicates a role, capacity, or function: *He serves as Department Head.* (One could say *serves like a Department Head*, referring to someone who is not a Department Head but serves *like* one.)

It is less desirable to use *like* as a conjunction to introduce a clause, although in formal writing this can be done if the clause is an elliptical one in which the verb is not expressed: *...looks like a good day for science.* Such a clause is not acceptable in formal writing if the verb is expressed: *...looks like it will be a good day.* In such a case, *like* must be replaced by *as*, *as if*, or *as though*: *...looks as if it will be a good day.* Fear of using *like* incorrectly may tempt a writer to use *as* when *like*, used as a preposition, is called for: *She acted like* (not *as*) *an idiot.*

may, might Originally, *might* was the past tense of *may*, but now both verbs are used as **subjunctives** expressing possibility or permission in present and future time. They differ in intensity rather than in time. In the sense of either possibility or permission, *may* is stronger than *might*: *He may go* makes a stronger case for his having permission to go or the likelihood that he will go than does *he might go.* (See **can, may**; *can* is used to express ability to do something.)

medium, media Like *datum/data*, *medium* is the singular form of *media*, but *media* is often used as a singular collective to refer to the means of mass communication taken as a whole. In careful writing, *media* as a subject should always be treated as a plural, and each individual means of mass communication should be expressed as the singular *medium*: *Television is an influential medium. Together, television, radio, newspapers, and periodicals make up the media.* Scientific writers should be careful to distinguish between these two forms of the noun when they are used in their original sense of a surrounding or pervading substance in which bodies exist or move: the environment. An object can only be surrounded by one *medium*, but different objects can be in different *media.* Bacteriologists and plant pathologists often incorrectly use *media* to refer to a single sterilized nutritive substance for cultivating bacteria or fungi. *Medium* is also used in other senses including that of an intervening thing through which a force acts or an effect is produced, or a person through whom communications come from the dead. *Medium* as an adjective also refers to something intermediate, a middle state or degree.

nor *Nor* is a conjunction used to express continuing negation; often it is paired with *neither*: *He was neither for the idea nor against it. She had no experience as a physiologist nor did the subject interest her.* The word *or* can be used instead of *nor* when the elements are within a single independent clause, and it is clear that the negative sense carries over to the element that is introduced: *He was not for or against the idea. She had no experience or interest in physiology.*

percent, percentage *Percent* (no longer written *per cent*) is specific and follows a number or numeral: *18 percent*. *Percentage* is nonspecific and should not be used with a number: *A small percentage died on day three.* (Use a space between a numeral and the percent sign: *18 %*. See Chapter 1.)

precede, proceed *Precede* means to come before in time, order, or rank or to introduce. *Proceed* means to go forward or onward, to undertake an action.

principal, principle *Principal* has several meanings both as a noun and as an adjective. As an adjective, it means leading or chief. As a noun, it means the leader or person in charge or the capital or main body of an estate or a sum owed as a debt: *The principal of the school is my pal.* *Principle* is used only as a noun meaning a basic truth, rule of human conduct, or fundamental law.

proved, proven Both are past participles of the verb *to prove*, but *proved* is preferred: *He had proved his point.* *Proven* is more common when used as an attributive adjective before a noun: *a proven record.* It is also the form used in the phrase *not proven.* (Because of variabilities caused by chance, and because of possible alternative explanations, scientists must be wary of saying that something is *proven.* We often strive to *disprove* a hypothesis.)

respectively The word means singly in the order designated. In scientific writing, it is often quite clear that the data presented in one brief list are given in the same order as the names or other information in another list, and the word *respectively* insults the intelligence of the reader: *Plants in the dark and in the light were etiolated and green, respectively.* With few exceptions, it is best to write the sentence so that *respectively* is not needed: *Proline concentrations were 0.5 mol*L^{-1} *for sample one, 1.32 mol*·L^{-1} *for sample two, and 3.56 mol*·L^{-1} *for sample three.* Don't write: *Proline concentrations were 0.5, 1.32, and 3.56 mol*·L^{-1} *for samples 1, 2, and 3, respectively.* The second way may be shorter, but it is inherently confusing.

shall, will In the most formal writing, use of these words was governed by a series of complex rules that would be understood by few modern readers. In modern usage, both are used to indicate futurity as well as determination, compulsion, or obligation, but *will* is becoming more common in construction of the future tense (opposite of the old rule): *Tomorrow, we will set up the experiment.* Because this can be ambiguous, it is a good idea to use some other construction when determination is the sense to be expressed: *Tomorrow, we must (have to, certainly will) set up the experiment.* *Shall* is sometimes used to emphasize resolve: *We shall overcome.* But *shall* is slowly disappearing from the language.

since, because *Since* is often used in the sense of *because*, which is still a good use. Some editors, however, might insist that its use be restricted to the following sense: *...the time that has elapsed* ***since*** *some event.*

so When the conjunction *so* introduces a clause that gives the purpose of, or reason for, an action stated earlier, it is usually followed by *that* and not preceded by a comma: *We changed the solutions so that toxins would not build up.* In that usage, it is a subordinating conjunction. *So* is used alone as a coordinating conjunction when the clause that it introduces states a result or consequence of something preceding: *We had to finish, so we worked late.*

so-called This hyphenated adjective precedes a noun that is not enclosed in quotation marks: *The dark period was interrupted by a so-called night-break.* Sometimes it is used sarcastically: *...a so-called leader.*

then, than Most commonly, *then* means *at that time* or refers to order or position: *First the seed germinates,* ***then*** *the seedling develops.* *Than* is used in comparisons: *Cultivar A is more drought hardy* ***than*** *cultivar B.*

that, which, who, whom See the discussion in Section 3.A on *Restrictive and Nonrestrictive Phrases and Clauses.* *That* (or *who* or *whom*) should be used to introduce restrictive phrases and clauses. *Which*, *who*, or *whom* (especially *which*), preceded by a comma, are used to introduce nonrestrictive phrases and clauses. *Who* or *whom* (depending on case) should be used when the reference is to people. *Which* is always used when it is preceded by *that* as a demonstrative pronoun: *We often long for that which is impossible.*

thus, thusly *Thus* is always the choice. Both are adverbs, and *thusly* never needs to be used.

toward, towards Either form is acceptable, but *toward* is most common in modern writing (especially in the United States).

unique Something *unique* is in a class by itself without an equal. As an adjective, it is an absolute. Therefore expressions of comparison should be avoided: *most unique*, *rather unique*, etc. But some modifiers are acceptable and logical: *nearly unique*, *more* (or *most*) *nearly unique*. The same is true of such other absolutes as **perfect** or **dead**.

were *Were* can be used as the past subjunctive mood of the verb *to be* to express conditions that are clearly hypothetical or contrary to fact: *If it were only possible.* Such statements may express a wish: *I wish that the data were complete.* Often they are preceded by *if*: *If our budget were only larger.* Informally, *was* is often used instead of *were*, and sometimes *were* is incorrectly used instead of *was* in indirect questions or when the conditional statement is not really contrary to fact: *...if the report was true, changes would be necessary. They asked if he was agreeable.*

9. SOME SUGGESTIONS ABOUT FORMAT AND WORD PROCESSORS

Most journals have printed instructions with many details on suitable format for manuscripts. The **CBE Style Manual Committee** (1994) also has an extensive discussion of format. Authors must carefully study such instructions before writing and certainly before submitting a manuscript to a particular journal. If the format does not follow that of the journal, reviewers may assume (sometimes correctly) that the manuscript was submitted to another journal and rejected. At the very least, an incorrect format tells the editor that the authors were too careless to bother checking such details before submitting a manuscript—and thus may be careless in carrying out the scientific study that is being described. Incorrect formatting also fails to consider the editors and reviewers who must read the manuscript.

A few suggestions, mostly my own personal preferences, come to mind: Manuscripts submitted for publication should *always* be double spaced; this includes captions, footnotes, quotations, *everything*. The double spacing is to help reviewers, copy editors, and compositors edit and otherwise mark the text. (Because it is now so easy to reproduce and even to rework a manuscript, it is quite in order for reviewers to write directly on the manuscript.) Titles should be both brief and descriptive. Because the title is usually the first thing a potential reader sees, its importance can't be overestimated. Next in importance is the abstract, which must also be as brief as possible (so as not to discourage a potential reader) while at the same time conveying all the key points including reasons for the study and the important conclusions. All plant material must be accurately named both in the abstract and in the sections describing methods and materials (see Chapter 2). Figures and tables convey the actual data produced by the study that is being reported, so it is extremely important to make them as complete and easy to understand as possible. (See the following section for many tips that can apply to published papers as well as to oral and poster presentations.) It is a common practice to use different symbols and lines for different treatments shown in figures and then to define these symbols and lines in the caption. Sometimes this is necessary because there may not be room on the figure to label the curves, but if the curves can be labelled on the figure, it becomes much easier for a reader to understand what is being presented. Why should an author play games with potential readers? Why should a reader have to learn a code to understand a figure? In any case, the caption should present enough information to make the figure understandable without having to read the text. Such details as dates and statistical treatments should also be included.

The preparation of manuscripts has been greatly aided by **word processing** programs that reduce the time required to create and especially to revise a manuscript and, most important, make it possible virtually to eliminate typographical and other errors. The spell checker is especially valuable, and grammar checkers can also help, especially those authors whose native language is not English. Such programs are now used by almost everyone world-wide, and many journals accept manuscripts in electronic form on a disc. Nevertheless, there are a few minor problems or irritations that result from the use of word processing programs.

There is a tendency to put too much faith in the word processor and not to proof the final manuscript before it is submitted. Perhaps the most common errors arise from editing the manuscript on the computer screen. One makes a change but may forget to remove all the material being replaced, for example. *It is still important to carefully proof the final document.*

The word processor usually provides a capability not previously enjoyed by most authors: the ability to justify the right margin (i.e., align it, as in this book). It seems that few authors are able to resist the temptation to use this capability although it sometimes produces some of those minor irritations. Although we are used to reading type-set material with justified right margins, there are two reasons why "amateur" justification of the right margin is not always a good idea:

First, many authors are reluctant to hyphenate (and it is somewhat easier for a typesetter at the printing press to work from a manuscript that has no hyphenation). The result with a justified right margin is that sometimes an exceptionally long word will not quite fit at the end of the line and is automatically moved (wrapped) by the word processor to the next line, leaving a large space that must be divided between the words that remain in the line. This produces a line with a few words separated by large spaces, like these, which can be distracting for a reader, who should be able to concentrate on the content of the paper. What does right justification gain for the editor and the reviewers? Is understanding the paper aided in any way by a justified right margin?

Second, some printers (especially dot-matrix) are not capable of dividing the space left at the end of a line into small fractions of a millimeter and proportioning it evenly among all the spaces between words and letters in the line. Instead, they work only with whole spaces (columns), and this means that spaces between some words are often at least one space larger than spaces between other words. This uneven spacing occurs especially with non-proportional fonts (e.g., Courier). Some readers also find this distracting.

If an author *must* justify the right margin of a manuscript (and it can make a good initial impression), great effort should be expended to hyphenate correctly (as professional typesetters have always done), and only the best of printers capable of microjustification should be used for the final product. Otherwise, right-margin justification gains nothing while it provides an added irritation that may not produce the desired receptive attitude in a reviewer.

10. SUMMARY

A. The sentence.

1. Two ideas in a sentence connected with a coordinating conjunction may share a common subject (or sometimes a verb), in which case they should not be separated with a comma or other punctuation: *The sentence presents one idea and adds another to fortify the first.*
2. Independent clauses connected by a coordinating conjunction should be separated by a comma before the conjunction: *The first clause has a subject and verb, and the second clause also has both subject and verb.*
3. Closely related sentences not connected with a conjunction may be separated by a semicolon; this ties the ideas together in a special way (as here). Separating such sentences with a comma is called a **comma fault** or **comma splice**, the habit must be avoided. (Did you notice the example?)
4. Less related sentences are separated by periods. These two sentences provide an example.
5. When two ideas are related to each other with a subordinating conjunction, they should be separated by a comma if the subordinate phrase or clause comes first in the sentence (*when* is the subordinating conjunction in this case); otherwise, no comma is needed.
6. And it is acceptable to begin a sentence with a coordinating conjunction (as here; used more as a conjunctive adverb) although this practice should not be overdone.

B. Modifying words.

1. Compound adjectives are formed by connecting with a hyphen: two adjectives, an adjective and a noun, or two nouns: *near-ultraviolet radiation, a ten-watt lamp, cell-wall structure.* (But omit the hyphen with numerals followed by unit symbols: 100 W lamp.)
2. In technical writing, it is important to use adverbial forms when a verb, adverb, or adjective is being modified: *an unusually concentrated solution.*
3. Nouns used as direct or indirect objects or following prepositions are always in the objective case, which in English is only evident when expressed by the personal pronouns *me*, *him*, *her*, *us*, *them*, and *whom* (or *whomever*).
4. Although some editors might disagree, authors would do well to use personal pronouns in writing technical articles for the scientific literature: *We homogenized the tissue in a buffer solution.* For one thing, this avoids use of the

passive voice: *The tissue was homogenized....* (Passive voice is appropriate for Methods and Materials sections.)

C. **Modifying phrases and clauses.**

1. A restrictive phrase or clause contains information that is essential to understand some part of the sentence not included in the phrase or clause; a nonrestrictive phrase or clause adds information that is not essential to understand the parts of the sentence not included in the phrase or clause.

   Nonrestrictive phrases or clauses are set apart by commas, but restrictive phrases or clauses are not: *The hypocotyl sections in auxin solution curved down. The hypocotyl sections, in auxin solution, curved down.* (The author determines the restrictiveness.)

   The relative pronoun *that* is used to introduce some restrictive phrases or clauses; it is not preceded by a comma: *This was the sample that we examined.*

   The relative pronoun *which* is used to introduce some nonrestrictive phrases or clauses; it is preceded by a comma: *We examined this sample, which* [incidently] *nearly escaped us.*

   Introductory or final nonrestrictive phrases or clauses are set off with a comma: *In preparation for the experiment, we collected the necessary glassware.*

   The relative pronoun *who* (or *whom*) is used with reference to people; *that* (or *which*) is used for everything else: *The teacher who drew the diagram.... The plant that grew....*

   Put two commas (a pair), or none, between subject and verb, or verb and object or subject complement: *The hypocotyl sections, in gibberellin solution, curved upward.* Not: *The hypocotyl sections in gibberellin solution, curved upward.*

2. Parenthetical phrases or clauses can be surrounded by commas, parentheses, brackets, or dashes. The choice is up to the author. If a phrase in parentheses comes at the end of a sentence, the period goes outside the parentheses. (Complete sentences initiated with the first word capitalized and terminated with a period can also be included in parentheses, like this.)

D. **Verbs.**

1. Verbs must agree in number (singular or plural) with their subjects.
2. Be consistent in verb tenses; use the past tense to describe methods and results.
3. Use the present and past participles correctly.
4. Whenever possible and appropriate, use the active instead of the passive voice.

E. **Further notes on punctuation.** This section contains some rules not contained in the previous discussion, which also often concern punctuation. Because the notes are already in a summary form, they are not repeated in this summary.

F. **Abbreviations.** Avoid using too many new abbreviations.

G. **Unnecessary words.** Tighten your writing by dropping unnecessary words and phrases and by using simple, concise forms whenever possible.

H. **Words with Special Problems.** Check the list here or use a good dictionary to use these words correctly.

I. Suggestions about format and word processors. Check instructions to authors published by the journal to which a manuscript is to be submitted. Watch for errors that tend to appear when one uses a word processor; proof the final printed manuscript carefully. Use a high quality printer and avoid justifying the right margin unless you use a suitable proportional font and use proper hyphenation at the end of lines that require it.

REFERENCES

Anonymous. 1993. The Chicago Manual of Style, Fourteenth Edition. The University of Chicago Press, Chicago and London.

de Mello Vianna, Fernando. 1977. The Written Word. Houghton Mifflin Co., Dictionary Division, Two Park Street, Boston, MA 02107

CBE Style Manual Committee. 1994. Scientific style and format: the CBE manual for authors, editors, and publishers. 6th edition. Cambridge University Press, Cambridge, New York. [See also earlier editions of CBE Style Manual.]

Heichel, G.H., D.E. Kissel, C.W. Stuber, G.A. Peterson, J.L. Hatfield, R.G. Hoeft, R.J. Wagenet, T.J. Logan, W.A. Anderson, and W.R. Luellen. 1990. Become a More Successful Author. Journal of the Soil Science Society of America. 54[Sept-Oct '90](5):iv-vii.

Pinckert, Robert C. 1986. Pinckert's Practical Grammar. Writer's Digest Books, 9933 Alliance Road, Cincinnati, Ohio 45242

Strunk, William, Jr. and E.B. White. 1979. The Elements of Style. The Macmillan Company, Toronto. Many writers say this is still the best guide for good writing.

CONSULTANTS

Ross E. Koning
Eastern Connecticut State University
Willimantic, Connecticut

Andrea L. Peterson
Utah State University
Logan, Utah

Shirlene M. Pope, Emeritus Professor
Utah State University
Logan, Utah

Moyle Q. Rice, Emeritus Professor
Utah State University
Logan, Utah

B

STANDARDS FOR EFFECTIVE PRESENTATIONS

Ross E. Koning
Biology Department
Eastern Connecticut State University
Willimantic, CT 06226-2295 U.S.A.

As scientists working in a rapidly advancing discipline, we must communicate effectively at regional, national, and international meetings. While publication is the permanent record of research progress, we rely heavily upon meetings presentations to communicate our most recent ideas and results. There are two common forms of meetings presentation: the oral report and the poster. This appendix is primarily designed to assist with the preparation of slides for the oral report, but much of the information applies equally well to artwork prepared for posters.

To communicate effectively, oral presentations must be designed to optimally deliver ideas and findings within a timed interval (usually less than 15 minutes). To achieve this, the artwork prepared for an oral report must be quite different from the artwork prepared for publication. Slides cannot have the fineness of detail nor the complexity required of publication graphics (for saving precious space in journals).

Since an oral report lacks the luxury of long explanations and detailed study, each slide (or viewgraph) must have a simple format, must be free of nonessential information, must be readily understood, and must have a single, clear purpose. Each slide must be visible, legible, attractive, and integrated with the other slides and the oral presentation. If properly designed, your graphics should catch the attention of the audience, reinforce your spoken ideas, and make your communication easier, faster, and more exciting.

Presentation planning software for computers may assist researchers in designing effective graphics, but because the power to design is still the prerogative of the scientist, some guidelines are needed to use this software to best advantage. Many examples of poorly designed slides generated by computer are evident at professional meetings. Thus, the guidelines presented here should be followed by those who use computer graphics facilities as well as those using older methods.

Clearly, slides must not detract from the presentation, so a key word for thinking about graphics is *simplify!* You do not want your audience to wallow in the particulars of your approach but to understand your results and to come to your conclusions (literally to *see* what you *mean*). It is important to keep methodological and statistical details from clouding the findings of your research. An interested party can ask questions about methods and analysis after your presentation or in a private conference. Your findings must stand above your process.

The following sections present some of my ideas about presentations.

1. SLIDE PRESENTATIONS

A. Planning your slides. To keep the audience listening and interested, you should organize and plan your presentation carefully before any artwork is attempted. The process is essentially the construction of a storyboard. Again, computer software has proliferated to assist business workers and (to a lesser extent) scientists in this process. First, compose your main message in 20 words or less. This should constitute your title. Assemble a sequence of similar short messages for each piece of evidence leading to your conclusion. Design a graphic that will communicate each element of the evidence *at a glance*. This might be a graph, a photograph, or a simple phrase. Allow only *one* idea per graphic! The audience must both look and listen, so it is critical to keep the slides and the spoken word simple and coordinated.

To help organize your presentation, a slide of your questions or main points might be projected near the start of the presentation. As you progress through your presentation, the **outline slide** could be shown again with a topic highlighted in a contrasting color to conclude the corresponding section or to introduce the next point in your presentation. At the end of the presentation, the list of questions or main points might be shown again with answers to reinforce your summary.

For complicated figures, use the build-up routine. Suppose you wish to show the difference between two curves sharing a common abscissa (x axis). In the first slide, the ordinate and abscissa are described and the first curve is shown. In the second slide, the additional ordinate is added and the second curve is shown. (The first curve and its ordinate may remain or may be drawn with thinner lines, or relegated to a different color in this slide.) In the third slide, the area between the two curves is hatched to emphasize the differences. In this way, three slides are used to present your findings. Now, to drive the trend home firmly, follow up with a text slide of a phrase boldly proclaiming the trend (*light stimulates shoot growth*). This final text graphic coupled with the build-up method makes your point clearly and memorably. Instead of dryly describing one graph with two lines on a single slide over a two-minute period, you will spend perhaps 30 seconds on each of four slides, each slide having a clear purpose. It helps your audience understand you better, and you will have held their attention to your presentation.

Not all graphics in your presentation need to hold information; **plain-color (no information) slides** draw attention to you and to important conclusions that you will simply state. This incredibly valuable and effective technique is seldom used but will really make your audience listen to what you are *saying*. Since the previous slide is replaced by plain color on the screen, the audience is forced to halt its examination and to listen to your interpretation.

Your planned presentation should have enough slides to prevent boredom for your audience. You should plan to change slides at 3-5 slides per minute (no more than 30 seconds spent on each slide). Limiting each slide to one idea should assure that you have enough slides. If you do not have enough slides planned to accommodate this change rate, then you probably have planned slides that present more than one idea! These need to be simplified. Walking into a 15-minute presentation with five complicated slides assures boredom for your audience and disrespect for you. This is particularly true for weary audiences after a few days of a national meeting. On the other hand, if you already have 50 slides for a 15-minute presentation, adequate rehearsal will help you decide if you cannot present all the slides (ideas) you may have planned. If, during rehearsal of your planned presentation, you cannot remember a particular point you want to make, you need another graphic element that will remind you of what was important and will help drive this point home to your audience as well.

Examine your presentation plan to be sure you have included **title slides, question slides, evidence slides,** and **conclusion slides.** Check carefully to note whether you need duplicate slides for graphics that are to be shown more than once in your presentation. The audience has no patience for you to give verbal instructions to a projectionist to try to locate a previous slide in a slide tray and then to return to another specific slide to continue your presentation. *Never turn back!* The storyboard must be unidirectional!

B. Preparing the artwork. Artwork must be designed to a 2-height-by-3-width ratio, which converts directly to the 24 x 36-mm format of the standard slide. It is best to keep the artwork away from the edges of the frame and to have central weight to the figure. Color and lettering weight can be used to emphasize. Realize that western audiences will view the slide from upper left to lower right, so items in the upper right and lower left corners of the diagram are of less importance and may go unnoticed in a complicated slide.

Legibility is a most-important criterion. **Make it bold!** Use large fonts! If your artwork output is on 8.5 x 11-inch or on A4 paper, tape it to a wall and stand three meters away. If you are using a computer screen to prepare artwork, again move three meters from the screen to check for legibility. At this distance, your graphic has the same visual size as a projection screen viewed from the back of a lecture room. Can you still read everything? Is the point of the graphic clearly observed from this distance? If not, then the lettering fonts, symbols, line thickness, or other elements of the artwork must be made larger or bolder.

As a final check for legibility, hold the graphic that has been converted to a 2 x 2-inch slide 40 cm from your nose. Check to be sure you can read everything and that the point of the graphic is clearly observed from this distance. If not, then either the lettering on the artwork must be proportionally larger or the slide must be retaken to more nearly fill the frame.

Legibility is assured if a few simple rules are applied to all graphics used in an oral presentation. A Kodak publication (Kodak #2, 1986) is a good technical discussion of legibility parameters and gives excellent tips for making your slides readable.

Text. Each textual graphic should be limited to a few (less than 10) words. A text slide should be thought of as an idea-gram. It makes your point dramatically and reminds you of what to say without having notes to handle (or shuffle!).

Complete sentences are seldom legible on a screen, so you should never have much to read on a slide. If you read a slide to the audience, you will bore them; if the audience reads a slide while you discuss it, they are not listening carefully to you. Helvetica or other sans-serif typefaces (such as this one) are much more readable in graphics than serif-fonts such as Times Roman (like this). The font size must be tall enough so that only 12 lines would fill the slide area from top to bottom, and wide enough so that only 36 characters would fill the slide area from left to right. The fonts must be **bold**; the thickness of the lines used to form each character should be between 1/10 to 1/5 of the height of the character. If these size guidelines are followed strictly, then everyone, even those in the back of the conference room, will be able to see what you intend to show them. These legibility rules apply equally to lettering on tables, graphs, and other artwork.

Tables. Prepare a slide showing *only* the portion of the table that you intend to discuss; leave out unused data. Excessive raw data and particularly unprocessed statistical tables are roadblocks between you and your audience; your audience is likely to be more interested in plant physiology than in statistical analysis. If you plan to focus only on a few cells in your table, then reformulate the table for projection. Reserve the complete table as a photocopy for private discussions with interested conferees. Alternatively, reserve the complete table as an **informational slide** after your final slide as potentially useful in answering a question from the audience. Moreover, if you wish to reach a larger audience, a handout might be desirable. However, if you do have a handout, pass it out only when the talk is *over*, so the audience will be listening and not reading during your presentation. Be sure your name, address, and phone number are printed on the handout so a member of the audience may contact you later for discussion.

In any case, tables for your oral presentation should have no more than five rows or columns to stay within the limits of legibility and sensibility. It is better to divide a large table into smaller portions on several slides than to present an illegible smear of tiny digits for several minutes. It is important to note that graphs (line, bar, pie) are usually *far* more illustrative and memorable than numbers in tables.

Line graphs should be bold and legible. Many computer graphics programs draw axis and plotted lines too finely by default and need to be modified for making bold projection slides. There should be no more than eight marked ticks on any axis, and all lettering should conform to the legibility rules for text (above). Again, computer programs usually default to small fonts suitable for publication but unsuitable for projection. Ideally, you should show only one or two lines or curves on each slide, but you may build up to a multiple-curve figure by revealing a new line on successive slides.

Bar graphs should have no more than eight bars and, if stacked bars are used, the bars may be (should be) revealed sequentially or in groups by color or texture for contrast.

Pie charts should have fewer than 10 slices. Color or texture contrast and "pulled-out" slices can be employed for emphasis. Sequential emphasis in a slide series may better communicate your findings.

Colors can be used to great advantage, but the overuse of colors is only distracting. The ease of applying color in computer-generated graphics tempts researchers to use too many colors. Whenever possible, associate a color with a particular type of information. The colors must contrast well in order to be distinguished. Light colors on a dark background are very legible. Dark colors on a medium background are seldom appreciated. Extreme contrast is needed for partially-darkened lecture halls, but in fully-darkened rooms extreme contrast fatigues the vision of your audience. High contrast black and white positives and negatives will be more easily appreciated if dyed to a pastel color. This takes the "edge" off of the excessive contrast. This is particularly important if you intend to mix black-and-white artwork with typical color slides or other continuous-tone artwork.

Do not underestimate the special connotations or emotional impact of certain colors (Xerox, 1985):

Red: stop, danger, fire, anger, warmth, passion, excitement
Blue: police, navy, sea, serenity, sky, fidelity, water, coolness
Green: go, growth, trees, country, spring, restfulness, youth, freshness, money (USA)
Yellow: caution, sunlight, cheerfulness, heat, light, life
White: hospitals, sterility, purity, innocence, peace, calm
Gray: somberness, dignity, quietness, age, wisdom, gravity

C. Making the slides. A variety of techniques can be employed to convert well-designed legible art work to effective slides. Scientists with access to computer graphics equipment can generate slides directly from their computers. Several highly sophisticated programs are now on the market to create beautiful, multicolored slides (e.g., Harvard Graphics, Power Point, SlideWrite, WordPerfect Presentations, etc.). It is possible to photograph these directly from the screen although resolution is not as good as when the computer file for the slide is sent to a film recorder designed to expose the film according to the file. Such recorders cost at least a thousand dollars (although prices have been dropping since they first came on the market). If you don't have the recorder, you can generate the presentation graphics on diskettes that are then sent to agencies to be converted to slides (at costs around $5.00 per slide.)

If the computer facilities are not available to you, or if you haven't time to go through the sometimes involved process of using computer programs to produce the slides, you can use the tried-and-true photographic processes that everyone had to use before the computer revolution. Basically, these processes involve putting the graphs, tables, words, or other materials on paper or other suitable medium and photographing the results to make slides. Such slides can be drawn by hand or produced with suitable software on a computer attached to a high-quality printer. Often, the University or other organization has a photography laboratory that will make the slides from the art work. Some workers like to or must go through the entire process themselves, using techniques described in the appendix to this appendix.

One quick and simple approach is to photograph the art work, on a good copy stand, using color slide film (direct positive). This gives fairly satisfactory results even with black-and-white copy, *providing that you remember to over expose by 1/2 to one full f stop* so that the background will appear bright instead of grey when projected. Of course, if you use color film, you can add color to the art work or even use color filters when making the slides.

H. Showing the slides. There is no replacement for adequate **rehearsal** of your slide presentation. It will show errors in your planning, errors in logic, errors in legibility, and certainly help you feel more confident when you actually deliver the oral report at the conference. Of course, this rehearsal must begin sufficiently before the conference so that errors can be corrected. You should begin to prepare your oral report as soon as possible.

During rehearsal be sure your slides are in mounts that will function in the standard Kodak Carousel 140 tray. Thin plastic mounts jam less frequently than any other type. The mounts should be numbered so that if a stack of slides is dropped, order can be quickly restored. You should place a dot or an x in the proper corner of the mount so that it will be oriented properly at the conference. The standard convention for placement of this mark is to hold the mount so that the slide can be properly read and to mark the lower left corner of the mount. The projectionist will then reorient the mount so the mark is in the upper right corner. This inversion of the slide will assure proper orientation on the screen. *Be sure to check this orientation* of your mark during rehearsal!

When you travel to the conference, carry your slides in a carry-on bag to avoid loss with your checked baggage. If your rehearsal has been adequate, you should arrive, slides in hand, at the conference confident that your presentation will be organized, legible, and understandable.

At the session for your presentation, arrive before the beginning of the session to present your (marked and ordered) slides to the projectionist and familiarize yourself with the room and its facilities. Bring photocopies and maybe an overhead transparency of each graphic just in case the slide projector fails. If the projector lamp burns out, your 15-minute time slot is too short to have the lamp replaced in time for you to finish with slides.

When it is your turn to speak, remember *never read a prepared speech*. There is no more boring method of delivery, and your audience unconsciously wonders if you did the work you are presenting. "If you did the work, then why must you read about it?" You should be able to discuss the artwork without a written text. Moreover, many auditoria are not equipped with reading lights. Let your artwork be your notes. *Never read a slide to the audience!* The only thing worse than projected sentences is projected paragraphs. Your audience thinks, "Why give an oral presentation; just publish the work!" Grammatically correct sentences with complicated logic and suitably condensed for journal publication are incomprehensible in an oral presentation; use simple, direct, conversational English.

Be careful in your use of a pointer, especially a light pointer (e.g., a laser pointer that projects a small red dot). It is very distracting to an audience when the light from the pointer dances or waves around excessively on the screen and especially when it dances all over the room while you are talking but not pointing. Hold it still or move it slowly to emphasize what you want, and *turn it off when it is not being used!*

I. Special language problems. You face an especially difficult challenge if you must give your presentation in a language that is not your native tongue. English has become the international language for scientific presentations, and most scientists everywhere now read and often speak English. Many scientists whose native language is not English, however, do not feel confident enough in the language to give an oral presentation without a written text, which breaks the rule just presented to *never read a prepared speech.* In spite of this rule, it is common at international meetings to hear presentations that are read in English by non-English-speaking scientists. The fact that the talk is read instead of being presented spontaneously is already enough to lose the interest of many if not most in the audience, but often the speaker is not highly proficient in English, so that his or her unfamiliar accent adds still another solid barrier between the speaker and the audience. There are several options available to you if you must present a talk in a language in which you are not highly competent. The following two options depend on the status of your presentation:

i. You arrive at the meeting with a conventional manuscript, perhaps one that has not even been edited by an English-speaking editor. You should strongly consider asking a fellow scientist whose native language is English (or the language of the conference) to read the paper for you. If you do, this person should have time to study the manuscript before presenting it, making grammatical and stylistic suggestions that will improve understanding. The reader should practice reading the manuscript and be admonished to read slowly with frequent pauses, even inserting spontaneous remarks that help to clarify what is being presented. Remember, however, that not all scientists whose native tongue is English are skillful readers who can read smoothly without backing up to correct small errors. If there is time, the friend might even study the manuscript until he or she has become so familiar with its contents that he or she can use your slides to present the material without reading it. Remember, you will be there to interrupt the person who is giving *your* talk; you can correct any mistakes or misconceptions—and you will be the one who answers questions during the discussion period. By having someone familiar with the language read your paper, you will almost certainly communicate with the audience better than if you read the paper yourself.

ii. You have time before going to the meeting to follow the suggestions that are presented in this section but still feel that you must read the paper. Remember that an oral presentation differs in significant ways from a published technical paper. Many of those who must present their talk in a foreign language simply read a manuscript that has been prepared exactly as if it were going to be published—as often it will be. Such a format is especially difficult for an audience to follow. Instead of using this rather standard procedure, outline your talk as a series of logically related, simple ideas, and prepare at least one slide for each of those ideas, as discussed above. Then write, probably in your own language to begin with so that ideas will flow freely, a short paragraph that describes each slide. Having done this, you can then either practice the English version until you are proficient—possibly so proficient that you won't have to read it after all—or you can persuade an English-speaking fellow scientist to read the paragraphs for you. You might even have time to send your slide-related manuscript to the friend before the meeting so that he or she can practice reading it. In any case, preparation based upon ideas and slides rather than the usual technical-manuscript format will lead to much better communication with your audience.

2. POSTER PRESENTATIONS

Part of the culture of modern plant science is that most presentations at professional meetings are now in the form of posters. The poster session consists of two distinct parts: the physical poster and informal discussions of the research work presented on the poster. Researchers prepare the physical poster at home but must also be prepared for the discussions to take place at the meeting.

A. The Physical Poster. Each society has specifications for their poster sessions, and these are usually found in the brochures calling for abstracts to be submitted for the meeting. The specifications provide the scientist with dimensions and orientation for the display boards. The entire poster must be prepared with the size and orientation in mind.

A four-foot-square poster (common in the U.S.; a little over a square meter) mounted with the top or center near eye-level is simple to prepare. The title goes at the top, and graphic elements placed virtually anywhere on the board will be visible at a glance. In general, a Western viewer will examine a poster from left to right and from top to bottom, so the elements are usually arranged to match. Contorted paths to follow the elements should be avoided.

If the specifications are for a 4x8-foot rectangle (or comparable metric dimensions), the preparations must be more careful. If the orientation is horizontal at eye level, the author must prepare the poster with the idea that the viewer will work across the poster from left-to-right. Moreover, with the general left-to-right movement of session participants through the poster displays, it will be difficult for discussants to have to move back to the left for a second row of graphic elements. Some meetings avoid this movement problem with 4x8-foot boards standing vertically. The reader will work down a poster with this orientation, but graphic elements placed above 6-feet from the floor or below waist-height are difficult to observe. These areas should be restricted to ancillary elements such as large photographs not needing much close examination. The essential elements of such a poster should be placed in a 4x4-foot square centered at eye level.

Bold, sans-serif fonts (like this) are preferred for all text. Fancy or calligraphic fonts should be avoided because of inherent poor legibility. The title of the poster should be lettered at 90-point or larger type. The title should state the major point or finding of the research clearly and in as few words as possible. In general it should match the title found in the published abstract and program booklet for the meeting. Nearby, the author(s) names and university or research organization and location should appear in moderate-sized type (perhaps 60-point). The major headings of the poster design should also be in 60-point type. The bulk of any text elements should be in 30-point type or larger.

The poster is not a "journal article on a board," and long passages of text are completely inappropriate for a poster session. Instead, the text elements should be largely **idea-grams** that lead the viewer through the other graphic elements to the conclusions drawn from the research. The sequence of presentation may be similar to a journal article, however.

A copy of the poster abstract might be the first graphic element after the title. The abstract puts the entire research body into a concise paragraph that a viewer can read to determine whether she or he wishes to continue to examine the poster or go on to something else. This might be the longest text element on the poster.

A brief introduction presenting the background of the research and perhaps introducing the particular organism studied might come next. This should be very brief and cover only the essentials. Three or four sentences would be a good guideline. A photograph could be appropriate here.

Unless the research details the development of a new procedure, an effective poster might simply include a flow chart rather than a long text explaining methods. In many cases a photograph or drawing communicates what would take many words to explain in text.

Most of the graphic elements on a poster present the research results. These graphic elements consist of photographs, graphs, tables, autoradiograms, etc. Effective posters frequently have simple figure captions or titles declaring the interpretation drawn from each accompanying graphic element. Each graphic element should be large and bold. The bounding rectangle for graphs and charts might be 20 x 30 cm. The minimum type size should be 30-point for all lettering. Graph symbols should approach 6 mm in width, and connecting lines should be 2 to 4 mm wide. The graphic elements can be slightly more complex than those presented in slides for oral presentations, but no more than three lines should normally appear on a graph (unless the figure presents a family of closely related curves), and bar charts should be limited to fewer than 10 bars. Each element should have a brief explanatory caption, but long passages are best avoided.

There is no discussion section on a poster. Never waste precious space on discussion text elements. The purpose of a poster session is to personally discuss your research with interested viewers. This interaction between scientists is the beauty of the poster session. The verbal discussions cover the details of the research and suggestions for improvements, etc. Your poster should provide the evidence and support for the verbal discussion. Of course, the main conclusions will appear in the abstract. This is important because some viewers will study your poster when you are not available for discussion.

The last graphic element in the poster sequence should be a summary of the research findings. A bulleted list is sometimes effective; a concluding mechanistic model diagram might be more memorable.

Very few viewers are prepared to write down a list of references to take home with them, so a literature-cited section is generally a waste of poster space. Established scientists will know the fundamental literature, and an interested newcomer will write to you after the meeting for references and reprints. Having a business card with you might be a handy and considerate alternative.

Some scientists bring summarizing handouts with them to give to interested parties. While these could be as formal as an article reprint, they are generally a "miniature" poster with a few literature references. It is difficult to judge how many copies of a handout to bring with you.

B. Setting up the poster. There are many styles of poster graphic elements. Careful use of color is effective; excessive use of color is distracting. Too many graphic elements are intimidating and stifle discussion. Many people bring the individual poster elements and their captions, etc. on individual cards or sheets, often on 8.5x11 or A4 sheets. These have to be mounted on the display boards individually; spacing and alignment must be adjusted at the meeting. This takes time that you might choose to spend in oral sessions or otherwise. A better plan is to attach the poster elements onto larger cardboards with spacing and alignment determined at home. Thus the assembly of the poster at the meeting involves only putting a small number of cardboards together. Of course, the smaller sheets may be easier to pack in your briefcase or suitcase.

C. The poster discussion. At the prescribed time, the author(s) are to stand at the poster and discuss the research with interested viewers. This is the opportunity to share ideas, to comment on techniques and interpretations, to improve the science, and sometimes to make new friends. The author should not approach each passer-by and launch into a deep presentation. Instead, the author lets the viewer initiate a discussion. The viewer usually is allowed to lead the discussion in a particular direction. On the other hand, occasionally a viewer will simply ask the author to "present" the research. Then the author is free to launch into a discussion in his/her own direction. Many viewers will simply want to examine the poster and draw their own unspoken conclusions without discussion.

Appendix to Appendix A

Photographic Techniques for Creating Slides

For those who lack computer photographic facilities or budgets for outside consultants, Kodak's pamphlet #3 (1987) and its additions (Kodak #4, 1982 and Kodak #5, 1982) are helpful as an overview of the process of making your own lecture slides with simple and relatively inexpensive photographic techniques. They present some of the options available to make reasonable slides for presentation. Another Kodak pamphlet (Kodak #6, 1985) is an excellent how-to booklet for several methods of producing text slides. A small poster (Kodak #7, 1987) reviews films (and their availability and processing), their uses, and techniques for particular applications (text, charts, line drawings, prints, electron micrographs, chromatograms, electrophoresis gels, autoradiograms, gross specimens, and lab scenes).

Your artwork will likely fall into two categories: continuous tone and high contrast.

Continuous tone artwork consists of images containing various shades of gray or various colors. These are best rendered into slides by photography with continuous tone black-and-white (e.g., Kodak Rapid Process Copy Film or reverse-processed Technical Pan Film) or color slide film (e.g., Ektachrome or Kodachrome).

High contrast artwork consists of images containing only black and white or vastly different shades of color. These are best rendered into slides by photography with high-contrast black-and-white films [e.g., Dektol-processed Kodalith, Technical Pan, or Precision Line (LPD4) films] or high-contrast color film [Vericolor Slide Film (SO-279)]. Kodalith, Technical Pan, and Vericolor films produce a "negative" slide of the artwork. The dark lines on the artwork are clear on the slide and the white background of the artwork will be intensely black (black-and-white films) or intensely colored (color determined by filter selection with the Vericolor film). Kodak Precision Line film produces a positive image (intensely black lines on clear background). As mentioned previously, high-contrast black-and-clear slides fatigue the vision of your audience in fully darkened rooms and should be dyed so that the clear areas take on a light pastel shade to reduce contrast.

D. Methods for continuous tone slides. Kodak Technical Pan Film 2415 is an incredibly fine-grained negative film that can be processed to various degrees of contrast (Kodak #8, 1982). It therefore makes a universal film for all sorts of applications. With a POTA developer [1.5 g 1-phenyl-3-pyrazolidinone (i.e., Phenodone), 30 g sodium sulfite per liter distilled water used for 15 min at 20 °C with agitation], normal contrast black and white negatives can be made for printing on photographic paper. Normal contrast black and white slides can be made by reversal processing as described in the box at the end of this appendix. Extremely high-contrast negatives for reversed-text slides or publication prints can be made by exposing the film at ASA 200 (1/30 s at f/11 using 4 150 W photofloods) and developing the film in undiluted Dektol for 2 min at 20 °C with continuous slow agitation. While the reversal processing could be used to make high-contrast, normal-text slides, LPD4 is a more convenient alternative.

As noted above, color transparency films (Ektachrome, Kodachrome, and equivalent) can be used to make slides from artwork. Color reproduction can be an advantage to distinguish portions of pie charts, bars in histograms, etc. However, standard black and white artwork loses some contrast with these films and appears dark gray on very light gray background. Moreover, any corrections and surface irregularities in the artwork will be visible in the final slide because of the low-contrast color rendition.

Color transparency films are very useful for showing the plant used, the methods and equipment employed, and your colleagues for the study, but tables, graphs, and line drawings from your data are better presented in slides of higher contrast.

E. Methods for high contrast slides. The best quality high-contrast negatives are prepared from Kodalith (or similar) graphic arts films processed in any of the Kodalith or similar (undiluted Dektol) developers. These materials give an intensely black background with very clear line-images. The film is insensitive to red safelight and, therefore, also to red or other faint-color guidelines on your artwork. Since the contrast of the film is so high, corrections to artwork made by clean erasure, white-out ink, and clean correction tape are invisible. Any undesirable marks that do appear can be blotted out on the negative with an opaque ink (e.g., from a permanent-black marking pen) or special opaquing material available in photo stores. The negative can be printed to make publication prints or it can be mounted in a slide mount for projection as a reversed-text slide. As a slide, the extreme contrast tempts one to dye the film prior to mounting. This is especially true when projection is to be in a small, completely darkened room. However, in large auditoria with incompletely draped windows or "house lights", the extreme contrast is highly desirable.

The very popular blue background slides are made with SO-279 Vericolor Slide Film (Kodak Pamphlet E-24). This is exposed for 6 seconds at f/16 through an O (range) "G" filter to artwork illuminated by 2 x 500 W Photofloods held 1 m away from and at a 45° angle to the artwork. The film is processed for 5 min at 35 °C in Unicolor K2 Chemistry (or equivalent C-41 processing). If you do not want to do color photographic processing, you can take the exposed film to a "One-Hour Photofinishing" company and ask for negatives only. These can then be mounted in Pakon (or equivalent thin-plastic) slide mounts. (The company might do the mounting.)

In a pinch, you may substitute Kodacolor II (or equivalent) for SO-279, expose it after metering at the manufacturer-recommended ASA, and have it processed for negatives only. The background colors will be weaker and the printed areas will be slightly orange.

LPD4 Kodak Precision Line Film is a direct-to-positive film for making black-line slides from black-line artwork. The film is exposed for 10 s at f/9.5 using 2 x 500 W Photofloods at 45° and 80 cm from copy center. It is developed 1 min in undiluted Dektol at 20 °C with slow continuous agitation. Fix, wash, and dry as any other film. The brilliant clear background and crisp black lines of these slides make them suitable for use in a large auditorium, for use in weak projectors, and for use in inadequately darkened rooms. These slides are excellent in any projection environment. A disadvantage is observed when these slides are projected in sequence with color slides or other less-brilliant slides. The contrast can be painfully excessive! If this is anticipated, the background can be dyed to reduce contrast. It is undesirable to reduce contrast very much, so very dilute solutions of water-soluble dyes should be used to obtain weak staining of the protein emulsion (Frost, T.M. and P.A. Jones, 1982).

F. Adding color to black-and-white slides. Negatives and slides from black-and-white films may be dyed in dilute solutions of water-soluble dyes. I suggest one-percent or more-dilute solutions of Tartrazine Yellow, Acid Orange II, or Napthol Green. Adding glacial acetic acid to make the dye solution 0.5 % acetic acid will improve the uniformity of the staining. A brief rinse in water after staining will prevent formation of opaque dye crystals on the film (Horner, J.A. and C. Pennington, 1974).

Individual lines of type can be emphasized, particularly on black-and-white reversed-text negatives, by highlighting them with water-soluble ink from felt markers (e.g., Vis á Vis™) designed for overhead projection. If the marks are made on the shiny (backing) side of the film, they can be washed off easily and reapplied as needed.

G. Recovering from disasters. Farmers Reducer is an amazing treatment that can eliminate the excess silver in the regions of a black-and-white slide or negative intended to be clear. An overexposed or overdeveloped negative slide can be corrected by cutting silver from the film. Farmers Reducer consists of two solutions: Part A (37.5 g potassium ferricyanide / 500 mL water) and Part B (240 g/L sodium thiosulfate). These have a reasonably long shelf-life, but mixtures of these two solutions are effective for less than 30 min and, therefore, must be prepared only immediately before use. The standard mixture is 1 part A with 4 parts B and 30 parts water. The slide, negative, or print is agitated in the solution for as long as it takes to remove the unwanted silver. If 30 min elapse before completion, fresh solution must be prepared and the process must be repeated. After sufficient clearing, wash the film in five volumes of water, dry, and mount. This simple treatment can save you from having to repeat the whole exposure and processing routine. Its availability also eliminates rationalizing the use of a single substandard slide from a roll of otherwise good frames.

There are intensifiers that might be helpful for correcting underexposure and underdevelopment, but I have not personally tried them. I guess I tend to err on the other side in an attempt to have very high contrast and the deepest possible black areas. The intensifiers can increase the contrast of thin, continuous tone negatives but may be less useful with slides.

Reversal Processing for Technical Pan Film

Use the following steps in the dark:

5 minutes 70 °C developer (D-19 undiluted stock solution)
5 volume water rinse
3 minute bleach (6 g $K_2Cr_2O_7$ + 7.5 mL Sulfuric Acid per 500 mL water)
5 volume water rinse
2 minutes clearing (50 g sodium sulfite per 500 mL water)

Use the following steps in the light:

5 volume water rinse
30 seconds per side of reel reexposure with 100 W bulb ca. 30 cm
1 minute Developer (D-19 as above)
5 volume water rinse

REFERENCES

Frost, T.M. and P.A. Jones. 1982. Do-it-yourself black and white slides. Bull. Eco. Soc. Amer. 63:16-17.

Horner, J.A. and C. Pennington. 1974. A simple and rapid method of adding color to photographic projection materials. Southeast Electron Microscopy Society Abstracts.

Kodak #1. 1975. Kodak Publication S-30, Planning and Producing Slide Programs.

Kodak #2. 1986. Kodak Publication S-24, Legibility: Artwork to Screen.

Kodak #3. 1987. Kodak Publication M3-106, Making lecture slides.

Kodak #4. 1982. Kodak Publication M3-515, Making lecture slides: Worksheet #1.

Kodak #5. 1982. Kodak Publication M3-516, Making lecture slides: Worksheet #2.

Kodak #6. 1985. Kodak Publication S-26, Reverse-text slides.

Kodak #7. 1987. Kodak Publication P-15, Kodak films for lecture slides.

Kodak #8. 1982. Kodak Publication P-255, Kodak Technical Pan Film 2415.

Kodak #9. 1984. Kodak Publication F-5, Kodak Professional Black and White Films.

Kodak #10. 1983. Kodak Publication G-73, Kodak Precision Line Films.

Xerox. 1985. Communicate effectively with slides. Reorder Number 610P153110. Xerox Reproduction Centers, Xerox Square, Rochester, NY 14644. [Very nice booklet puts much of this appendix down in an outline with color examples.]

CONSULTANT

Frank B. Salisbury
Utah State University
Logan, Utah

C

GUIDELINES FOR MEASURING AND REPORTING ENVIRONMENTAL PARAMETERS FOR PLANT EXPERIMENTS IN GROWTH CHAMBERS

Developed by the American Society of Agricultural Engineers Environment of Plant Structures Committee; approved by the ASAE Structures and Environment Standards Committee; adopted by ASAE, March 1982. Revised March 1986; reconfirmed December 1989; revised February 1992[3].

Submitted by:

John C. Sager
NASA, John F. Kennedy Space Center
KSC, FL, 32899-0001 U.S.A.

Donald T. Krizek
USDA Climate Stress Laboratory
U. S. Department of Agriculture, ARS,
Beltsville, MD 20705-2350 U.S.A.

Theodore W. Tibbitts
Department of Horticulture
University of Madison
Madison, WI 53706-1590 U.S.A.

SECTION 1: PURPOSE AND SCOPE

1.1 The purpose of this Engineering Practice is to set forth guidelines for the measurement of environmental parameters that characterize the aerial and root environment in a plant growth chamber.

1.2 This Engineering Practice establishes criteria that will promote a common basis for environmental measurements for the research community and the commercial plant producer.

1.3 This Engineering Practice promotes uniformity and accuracy in reporting data and results in the course of conducting plant experiments.

---

[3]Submitted July 16, 1993; includes a few recent modifications. This is ASAE Engineering Practice: ASAE EP 411.2

SECTION 2: INTRODUCTION

2.1 The aerial environment is characterized by the following parameters: air temperature, atmospheric composition including moisture and carbon dioxide concentration, air velocity, radiation, and the edge effects of wall/floor on these parameters.

2.2 The root environment is characterized by the following parameters: medium composition and quantity, nutrient concentrations, water content, temperature, *p*H, electrical conductivity, and oxygen concentration.

2.3 Measuring and reporting these various parameters will be covered in the sections that follow. The definitions of the parameters indicate the symbol and units in the format (symbol, units). Measurements should be made that accurately represent the mean and range of the environmental parameters to which the plants are exposed during the experimental period, to indicate the temporal variations, both cyclic and transient, and the spatial variations over the separate plants in the chamber.

2.4 The definitions, measurement techniques, and reporting procedures provide criteria and promote uniformity in measuring and reporting environmental parameters, but these guidelines should not be used to select the environmental parameters applicable to a particular experiment. Other parameters may be applicable to a particular experiment or special environments such as elemental concentration in hydroponic solutions, pollutant concentration in air quality research, and spectral quality ratios in photobiology.

2.5 When measurements are made, the chamber should be operating with containers and plants located in the chamber. Provision should be made to take all measurements with minimum disturbance to the operating environment.

SECTION 3: DEFINITIONS

3.1 **Radiation**: The emission and propagation of electromagnetic waves or particles through space or matter.

3.1.1 **Radiant energy** (Q_e, J): The transfer of energy of radiation.

3.1.2 **Energy flow rate** (ϕ_e, W): The rate of flow of energy, a fundamental radiometric unit; also called **radiant power.**

3.1.3 **Spectral distribution:** A functional or graphic expression of the relation between the spectral energy flux, spectral photon flux, or fluence rate per unit wavelength, and wavelength.

3.1.4 **Spectral energy flow rate** ($\phi_{e,\lambda}$, $W \cdot nm^{-1}$): The radiant energy flow rate per unit wavelength interval at wavelength λ.

3.1.5 **Energy flux** (E_e, $W \cdot m^{-2}$): The radiant energy flow rate per unit plane (flat) surface area; also called irradiance.

3.1.6 **Spectral energy flux** ($E_{e,\lambda}$, $W \cdot m^{-2} \cdot nm^{-1}$): The radiant energy flow rate per unit plane surface per unit wavelength interval at wavelength λ.

3.1.7 **Energy fluence** (F_e, $J \cdot m^{-2}$): The radiant energy dose time integral per unit spherical area.

3.1.8 **Spectral energy fluence** ($F_{e,\lambda}$, $J \cdot m^{-2} \cdot nm^{-1}$): The energy fluence per unit wavelength interval at wavelength λ.

3.1.9 **Energy fluence rate** ($F_{e,t}$, $W \cdot m^{-2}$): The radiant energy fluence per unit time. The same as radiant energy flux (irradiance) for normal incident (perpendicular) radiation on a plane surface.

3.1.10 **Spectral energy fluence rate** ($F_{e,t,\lambda}$, $W \cdot m^{-2} \cdot nm^{-1}$): The radiant energy fluence rate per unit wavelength interval at wavelength λ.

3.1.11 **Photon** (unit = q; i.e., one photon): A quantum (the smallest, discrete particle) of electromagnetic energy with an energy of hc/λ (h = Planck's constant; c = speed of light; λ = wavelength). Its energy is expressed in joules (J).

3.1.12 **Photon flow rate** (ϕ_p, $q \cdot s^{-1}$ or $mol \cdot s^{-1}$): The rate of flow of photons.

3.1.13 **Photon flux** (E_p, $q \cdot m^{-2} \cdot s^{-1}$ or $mol \cdot m^{-2} \cdot s^{-1}$): The photon flow rate per unit plane surface area; sometimes also called photon flux density to emphasize the unit area.

3.1.14 **Spectral photon flux** ($E_{p,\lambda}$, $q \cdot m^{-2} \cdot s^{-1} \cdot nm^{-1}$ or $mol \cdot m^{-2} \cdot s^{-1} \cdot nm^{-1}$): The photon flux per unit wavelength interval at wavelength λ.

3.1.15 **Photon fluence** (F_p, $q \cdot m^{-2}$ or $mol \cdot m^{-2}$): The photon flow rate per unit spherical area.

3.1.16 **Photon fluence rate** ($F_{p,t}$, $q \cdot m^{-2} \cdot s^{-1}$ or $mol \cdot m^{-2} \cdot s^{-1}$): The photon fluence per unit time. The same as photon flux for normal incident radiation.

3.1.17 **Spectral photon fluence rate** ($F_{p,t,\lambda}$, $q \cdot m^{-2} \cdot s^{-1} \cdot nm^{-1}$ or $mol \cdot m^{-2} \cdot s^{-1} \cdot nm^{-1}$): The photon fluence rate per unit wavelength interval at wavelength λ.

3.1.18 **Light**: Visually evaluated radiant energy, with wavelengths approximately ranging between 380 and 780 nm, based on sensitivity of the human eye.

3.1.19 **Illuminance** (E_v, lx): The luminous flux (light incident per unit area).

> NOTE: (a) Radiation instruments that measure illuminance are not recommended. They should only be used along with recommended radiation instruments for historical comparison. (b) Conversion factors from illuminance to radiation are spectrally sensitive and thus unique for each specified source.

3.1.20 **Photosynthetically active radiation** (*PAR*, $q \cdot m^{-2} \cdot s^{-1}$, $mol \cdot q \cdot m^{-2} \cdot s^{-1}$, or $W \cdot m^{-2}$): The radiation in the wavelength range of 400-700 nm. Measured as the photosynthetic photon flux (*PPF*), in $quanta \cdot m^{-2} \cdot s^{-1}$, or $mol \cdot m^{-2} \cdot s^{-1}$ or photosynthetic irradiance (*PI*) in $W \cdot m^{-2}$ for the specified waveband, λ_1-λ_2 (400-700 nm).

3.1.21 **Photomorphogenic radiation** ($q \cdot m^{-2} \cdot s^{-1}$, $mol \cdot m^{-2} \cdot s^{-1}$, or $W \cdot m^{-2}$): The radiation with wavelengths approximately ranging between 300-800 nm contributing to photomorphogenic responses (i.e., phototropism, flowering, reproduction, elongation, dormancy) in relation to the relative quantum efficiency of the spectral quality of the radiation in several discrete spectral regions. Measured as the photon flux in average quanta·m^{-2}·s^{-1}, or in energy flux in W·m^{-2} for the specified waveband, λ_1-λ_2.

NOTE: The specfic responses to photomorphogenic radiation must be biologically quantified and carefully measured for each **response spectrum (action spectrum)**.

3.2 **Temperature:** The thermal state of matter with reference to its tendency to transfer heat. A measure of the mean molelcular kinetic energy of that matter.

3.2.1 **Temperature, dry bulb** (T, °C): The temperature of a gas or mixture of gases indicated by an accurate thermometer protected from or corrected for radiation effects.

3.2.2 **Temperature, wet-bulb** (T_w, °C): Wet-bulb temperature is the temperature indicated by a wet-bulb sensor of a psychrometer constructed and used according to instructions.

3.2.3 **Temperature, dewpoint** (T_d, °C): The temperature of an air mass at which the condensation of water vapor begins as the temperature of the air mass is reduced. Also, the temperature corresponding to saturation vapor pressure (100 % relative humidity) for a given air mass at constant pressure.

3.3 **Atmospheric moisture:** The water vapor component of the mixture of gases of the atmosphere.

3.3.1 **Water vapor density** (P_r, g·m^{-3} or Pa): The ratio of the mass of water vapor to a given volume of air, also called absolute humidity. It may also be measured as partial pressure. (**Water vapor pressure**).

3.3.2 **Relative humidity** (H_r, percent): The ratio of the mole fraction of water vapor present in the air to the mole fraction of water vapor present in saturated air at the same temperature and barometric pressure. It approximates the ratio of the partial pressure or density of the water vapor in the air to the saturation pressure or density of water vapor at the same temperature.

3.3.3. **Water vapor deficit** (e_d, Pa): The difference between saturation water vapor pressure at ambient temperature and actual vapor pressure at ambient temperature.

3.4 **Air velocity** (V, m·s^{-1}): The time rate of air motion along a directional vector.

3.5 **Carbon dioxide concentration** ([CO_2], μmol·mol^{-1} or Pa): The carbon dioxide component of the mixture of gases of the atmosphere. Current expression of units of equivalent gas concentration are μmol·mol^{-1}, parts per million (ppm), or μL·L^{-1}, but they do not express standard temperature and pressure, *STP*, correction. Use of partial pressure, Pa, is preferred in nonstandard atmospheres.

3.6 **Watering** (volume, L): The addition of water to the substrate specified as to the source, the times, the amount, and the distribution method.

3.7 **Substrate:** The media comprising the root environment specified as to type, amendments, and its dimensions (container size).

3.8 **Nutrition:** The organic and inorganic nutrient salts necessary for plant growth and development. Formula and/or macro and micro nutrients are specified within the substrate as $mol \cdot m^{-3}$ or within liquid solution as $mol \cdot L^{-1}$.

3.9 **Hydrogen ion concentration** (*p*H units): The hydrogen ion concentration measured in the substrate or liquid media over a range of 0 to 14 *p*H units.

3.10 **Electrical conductivity** (λ_c, $mS \cdot m^{-1}$): The electrical conductivity within the solid or liquid medium.

3.11 **Accuracy:** The extent to which the readings of a measurement approach the true values of a single measured quantity.

3.12 **Precision:** The ability of the instrument to consistently reproduce a value of a measured quantity.

SECTION 4: INSTRUMENTATION

4.1 **Radiation.** Sensors should be cosine corrected and constructed of material of known stability, known response curve, and low temperature sensitivity. Such relationships should be specified and available for each sensor. By definition fluence measurements can only be taken with spherical sensors and cannot be derived from measurements taken with any plane-surface sensors. The sensitivity and linearity over the spectral response and irradiance range should be specified by calibration or direct transfer from a calibrated instrument. Spectral measurements should be made with a bandwidth of 20 nm or less in the 300-800 nm waveband.

4.2 **Temperature.** Sensors should be shielded with reflective material and aspirated ($\geq 3\ m \cdot s^{-1}$) for air measurements.

4.3 **Atmospheric moisture.** Measurement should be made by infrared analyzer, dewpoint sensor, or psychrometer (shielded and aspirated at $\geq 3\ m \cdot s^{-1}$).

4.4 **Air velocity.** Sensor should have a range of 0.1 to 5.0 $m \cdot s^{-1}$.

4.5 **Carbon dioxide.** Measurement should be made by an infrared analyzer with a range of 0 to 1000 $\mu mol \cdot mol^{-1}$ or greater.

4.6 **Hydrogen ion concentration.** Sensor should have a range of 3.0 to 10.0 *p*H units.

4.7 **Electrical conductivity.** Sensor should have a range of 1 to 10^{-2} $mS \cdot m^{-1}$ (1-100 milliohms resistance).

4.8 **Expected instrument precision and measurement accuracy.** Table 1 gives these percentages, which indicate full scale precision or accuracy. Further definition of these requirements can be found in reference 27.

Table 1. Expected Instrument Precision and Measurement Accuracy

Parameter	Instrument precision	Measurement accuracy of reading
Radiation		
Flux	± 1 %	± 10 %
Spectral flux	± 1 %	± 5 %
Temperature		
Air	± 0.1 °C	± 0.2 °C
Soil or liquid	± 0.1 °C	± 0.2 °C
Atmospheric moisture		
Relative humidity	± 2 %	± 5 %
Dewpoint temperature	± 0.1 °C	± 0.5 °C
Water vapor density	± 0.1 $g \cdot m^{-3}$	± 0.1 $g \cdot m^{-3}$
Air velocity	± 2 %	± 5 %
Carbon dioxide	± 1 %	± 3 %
*p*H		
H^+ concentration	± 0.1 *p*H	± 0.1 *p*H
Electrical conductivity		
Salt concentration	± 5 %	± 5 %

SECTION 5: MEASUREMENT TECHNIQUE

5.1 **Photon and energy flux.** Measurements should be taken over the top of the plant canopy to obtain the average, maximum, and minimum readings, and at least at the start and end of each study and biweekly if studies extend beyond 14 days.

5.2 **Spectral photon or energy flux.** A measurement should be taken at the center of the growing area, at least at the start and end of each study.

5.3 **Air temperature.** Measurements should be made at the top of the plant canopy at least daily, 1 h or more after each light and dark period begins, to obtain average, maximum, and minimum data. Continuous measurements are recommended.

5.4 **Soil and liquid temperatures.** Measurements should be made at the center of the containers in the growing area, obtaining average, maximum, and minimum readings at the middle of the light and dark periods at the start of the experiment. Continuous measurements during the entire study are recommended.

5.5 **Atmospheric moisture.** Measurements should be made at the top of the plant canopy in the center of the growing area daily, 1 h or more after each light and dark period. Continuous measurements are recommended.

5.6 **Air velocity.** Measurements should be taken at the top of the plant canopy, at the start and end of the studies. Obtain average, maximum, and minimum readings over the plants. If instantaneous devices are utilized, 10 consecutive readings should be taken at each location and averaged.

5.7 **Carbon dioxide.** Measurements should be taken at the top of the plant canopy continuously during the period of the study. A time-sharing technique that provides a periodic measurement (at least hourly) in each chamber can be utilized.

5.8 **Watering.** The quantity of water added to each container or average per plant at each watering should be measured. Soil moisture should be measured to provide the range between waterings.

5.9 **Nutrition.** Measurement of nutrients added to a volume of medium or concentration of nutrients added in liquid culture should be obtained at each addition.

5.10 **Hydrogen ion concentration.** The *p*H of the liquid solutions in a nutrient culture system should be monitored daily and before each *p*H adjustment. The *p*H of the solution extracted from solid media should be measured at the start and end of studies and before and after each *p*H adjustment.

5.11 **Electrical conductivity.** Conductivity of the liquid solutions in a nutrient culture system should be monitored daily during the course of each study. Conductivity of the solution extracted from solid media should be measured at the start and end of each study.

SECTION 6: REPORTING

6.1 **Photon or energy flux.** Report the average and range over the containers at the start of the study, and the decrease or fluctuations from the average over the course of the study. The source of radiation and the measuring instrument/sensor should be reported. Illuminance should not be reported except for historical comparison in conjunction with other radiation measurements.

6.2 **Spectral photon or energy flux.** Report the spectral distribution (graphical) and the integral (photon or energy flux) at the start of the study. The source of radiation and the measuring instruments should be reported.

6.3 **Air temperature.** Report the average daily readings with extremes over the growing area for the light and dark periods with the range of variations over the course of the study.

6.4 **Soil and liquid temperatures.** Report the average readings at the start of the study for the light and dark periods.

6.5 **Atmospheric moisture.** Report the daily average moisture level for both light and dark periods with the range over the course of the study.

6.6 **Air velocity.** Report the average and range over containers at the start and end of the study.

6.7 **Carbon dioxide.** Report the mean of hourly average concentrations and range of average readings over the period of the study.

6.8 **Watering.** Report the frequency of watering, source, and amount of water added daily to each container, and/or the range in soil moisture content between waterings.

6.9 **Substrate.** Report the type of soil and amendments, or components of soilless substrate, and container dimensions.

6.10 **Nutrition.** Report the nutrients added to solid media. Report the concentration of nutrients in liquid additions and in liquid culture solution along with the amount and frequency of all additions.

6.11 **Hydrogen ion concentration.** Report the mode and range in *p*H over the period of the study.

6.12 **Electrical conductivity.** Report the average and range in conductivity over the period of the study.

SECTION 7: SYNOPTIC TABLE

7.1 Table 2 is a synoptic table of the material presented in the previous section.

Table 2. Guidelines for Measuring and Reporting Environmental Parameters for Plant Experiments in Growth Chambers*

		Measurements		
Parameter	**Units**[a]	**Where to take**	**When to take**	**What to report**
Radiation **Photon flux** λ_1 - λ_2 nm, with cosine correction or **Energy flux (Irradiance)**[b], λ_1 - λ_2 nm with cosine correction	$\mu mol \cdot m^{-2} \cdot s^{-1}$ (λ_1 - λ_2 nm) or $W \cdot m^{-2}$ (λ_1 - λ_2 nm)	At top of plant canopy. Obtain maximum and minimum over plant growing area.	Minimum measurements: at start and finish of each study and biweekly if studies extend beyond 14 d.	Average (± extremes) over containers at start of study. Percent decrease or fluctuation from average over the course of the study. Source of radiation and instrument/sensor.

Continued

Spectral photon flux $\lambda_1 - \lambda_2$ nm, in <20 nm bandwidths with cosine correction or **Spectral energy flux (Spectral irradiance)** $\lambda_1 - \lambda_2$ nm, in <10 nm bandwidths with cosine correction	$\mu mol \cdot m^{-2} \cdot s^{-1} \cdot nm^{-1}$ ($\lambda_1 - \lambda_2$ nm) or $W \cdot m^{-2} \cdot nm^{-1}$ ($\lambda_1 - \lambda_2$ nm)	At top of plant in center of growing area.	Minimum measurement: at start and end of each study.	Spectral distribution of radiation with integral ($\lambda_1 - \lambda_2$) at start of study. Source of radiation and instrument/sensor.
Photosynthetic photon flux, *PPF*,[c] $\lambda_{400} - \lambda_{700}$ nm with cosine correction or **Photosynthetic irradiance, *PI*,**[c] $\lambda_{400} - \lambda_{700}$ nm with cosine correction	$\mu mol \cdot m^{-2} \cdot s^{-1}$ ($\lambda_{400} - \lambda_{700}$ nm) or $W \cdot m^{-2}$ ($\lambda_{400} - \lambda_{700}$ nm)	At top of plant canopy. Obtain maximum and minimum over plant growing area.	Minimum measurement: at start and finish of each study and biweekly if studies extend beyond 14 d.	Average (± extremes) over containers at start of study. Percent decrease or fluctuation from average over the course of the study. Source of radiation and instrument/sensor.
Temperature Air Shielded and aspirated ($\geq 3\ m \cdot s^{-1}$) device	°C	At top of plant canopy. Obtain maximum and minimum over plant growing area.	Minimum measurement: measure once daily during each light and dark period at least 1 h after light change. Desirable: continuous measurement.	Average of once daily readings (or hourly average values) for the light and dark periods of the study with ± extremes for the variation over the growing area.
Temperature Soil and liquid	°C	In center of container. Obtain maximum and minimum over plant growing area.	Minimum: measure at the middle of the light and dark periods at the start of the study. Desirable: continuous measurement.	Light and dark period readings at the start of the study (or hourly average values of 24 h if taken).

Continued

Table 2. Guidelines for Measuring and Reporting Environmental Parameters (continued)

Parameter	Units[a]	Measurements		
		Where to take	When to take	What to report
Atmospheric Moisture Relative humidity (RH) with aspirated psychrometer, dewpoint hygrometer, or IRGA or Vapor deficit, *VPD* or vapor difference	% RH, dewpoint temperature, or $g \cdot m^{-3}$ or kPa or $g \cdot m^{-3}$	At top of plant canopy in center of plant growing area.	Minimum: once during each light and dark period at least 1 h after light changes. Desirable: continuous measurement.	Average of daily readings for both light and dark periods, with range of daily variation during studies.
Air Velocity	$m \cdot s^{-1}$	At top of plant canopy. Obtain maximum and minimum readings over growing area.	At start and end of studies. Take 10 successive readings at each location and age.	Average reading and range over containers at start and end of the study.
Carbon Dioxide Mole fraction Partial pressure Concentration	$\mu mol \cdot mol^{-1}$ Pa $mol \cdot m^{-3}$	At top of plant canopy	Minimum: hourly measurements. Desirable: continuous measurements.	Mean of hourly average concentrations and range of average concentrations over the period of the study.
Watering	liter (L)		At times of water additions.	Frequency of watering. Amount of water added and/or range in soil moisture content between waterings.
Substrate			At beginning of studies.	Type of soil and amendments. Components of soilless substrate. Water retention capacity. Container dimensions.

Continued

Nutrition	Soil media $mol \cdot m^{-3}$ or $mol \cdot kg^{-1}$ Liquid culture $mol \cdot L^{-1}$		At times of nutrient additions.	Nutrients added to solid media. Concentration of nutrients in liquid additions and solution culture. Amount and frequency of solution addition and renewal.
***p*H**	*p*H units	In saturated media, extract from media or in solution of liquid culture.	Start and end of studies in solid media. Daily in liquid culture. Before each *p*H adjustment.	Mode and range during studies.
Electrical conductivity	$mS \cdot m^{-1}$[d] (millisiemens per meter)	In saturated media, extract from media or in solution of liquid.	Start and end of studies in solid media. Daily in liquid culture.	Average and range during studies.

*USDA North Central Regional (NCR 101) Committee on Controlled Environment Technology and Use, June 1978; Revised by ASAE Environment of Plant Structures Committee, Oct. 1978; Revised by NCR 101 Committee, March 1993. Published in part in the following references: 1, 17, 18, 22, 27, 28, 34, and 37.

[a] Report in other subdivisions of indicated units if more convenient.

[b] The energy flux (irradiance) is also commonly reported in $J \cdot m^{-2} \cdot s^{-1}$ (equals $W \cdot m^{-2}$).

[c] Referred to as photosynthetically active radiation (*PAR*) for general usage.

[d] $mS \cdot m^{-1} = 10\ \mu mho \cdot cm^{-1}$.

BIBLIOGRAPHY

American Society for Horticultural Science Working Group on Growth Chambers and Controlled Environments. 1980. Guidelines for measuring and reporting the environment for plant studies. HortScience 15(6):719-720.

ASAE. 1988. ASAE Engineering Practice: ASAE EP285.7, Use of SI (Metric) Units. American Society of Agricultural Engineers, St. Joseph, MI 49085.

ASAE. 1985. ASAE Engineering Practice: ASAE EP402, Radiation Quantities and Units. American Society of Agricultural Engineers, St. Joseph, MI 49085.

Bell, C.J. and D.A. Rose. 1981. Light measurement and the terminology of flow. Plant, Cell and Environment 4:89-96.

Bickford, E.D. and S. Dunn. 1972. Lighting for Plant Growth. The Kent State University Press, Kent, Ohio.

Biggs, W.W. and M.C. Hansen. 1979. Instrumentation for biological and environmental sciences. LI-COR Inc. Lincoln, Nebraska.

C.I.E. 1970. International Lighting Vocabulary. Publ. No. 17. *Commission Internationale de l'Eclairage*, Paris, France.

Downs, R.J. 1975. Controlled Environments for Plant Research. Columbia University Press, New York, NY.

Downs, R.J. 1988. Rules for using the international systems of units. HortScience 12(5):811-812.

Geist, J. and E. Zalewski. 1973. Chinese nomenclature for radiometry. Applied Optics 12:435-436.

Holmes, M.G. and L. Fukshansky. 1979. Phytochrome photoequilibrium in green leaves under polychromatic radiation: a theoretical approach. Plant, Cell and Environment 2:59-65.

Holmes, M.G., W.H. Klein and J.C. Sager. 1985. Photons, flux, and some light on philology. HortScience 20(1):29-31.

IES Lighting Handbook. 1981. Fifth edition. Illuminating Engineering Society of North America, New York, NY.

Incoll, L.D., S.P. Long and M.R. Ashmore. 1977. SI units in publications in plant science. Current Advances in Plant Science 9(4):331-343.

Kerr, J.P., G.W. Thurtell and C.B. Tanner. 1967. An integrating pyranometer for climatological observer stations and mesoscale networks. Journal of Applied Meterology 6:688-694.

Kozlowski, T.T., editor. 1968 and 1976. Water Deficits and Plant Growth. Vol. 1, Development, Control and Measurements. Vol. 4, Soil Water Measurement, Plant Responses and Breeding for Drought Resistance. Academic Press, New York, NY.

Krizek, D.T. 1982. Guidelines for measuring and reporting environmental conditions in controlled-environment studies. Physiol. Plant. 56:231-235.

Krizek, D.T. and J.C. McFarlane. 1983. Controlled-environment guidelines. HortScience 18(5):662-664 and Erratum 19(1):17.

Langhans, R.W., editor. 1978. A Growth Chamber Manual. Cornell University Press, Ithaca, New York.

LI-COR. 1982. Radiation measurements and instrumentation. Publ. No. 8208-LM. LI-COR, Lincoln, Nebraska.

McCree, K.J. 1972. Test of current definitions of photosynthetically active radiation against leaf photosynthesis data. Agricultural Meteorology 10:443-453.

McFarlane, J.C. 1981. Measurement and reporting guidelines for plant growth chamber environments. Plant Science Bulletin 27(2):9-11.

Mohr, H. and E. Schäfer. 1979. Guest Editorial—Uniform terminology for radiation: A critical comment. Photochemistry and Photobiology 29:1061-1062.

Monteith, J.L. 1984. Consistency and convenience in the choice of units for agricultural science. Experimental Agriculture 20(2):105-117.

NBS Technical Note 910-2. 1978. Self-Study Manual on Optical Radiation Measurements, Part 1—Concepts. United States Government Printing Office, Washington, DC.

Norris, K.H. 1968. Evaluation of visible radiation for plant growth. Annual Review of Plant Physiology 19:490-499.

North Central Regional 101 Committee on Growth Chamber Use. 1986. Quality assurance procedures for environmental control and monitoring in plant growth facilities. Biotronics 15:81-84.

Percival Manufacturing Co. 1981. Guidelines: Measuring and Reporting Environment for Plant Studies. Available as a plastic card.

Rosenberg, N.J. 1974. Microclimate: The Biological Environment. John Wiley & Sons, New York, NY.

Rupert, C.S. and R. Latarjet. 1978. Toward a nomenclature and dosimetric scheme applicable to all radiations. Photochemistry and Photobiology 28:3-5.

Salisbury, F.B. and C.W. Ross. 1991. Plant Physiology. Fourth edition. Wadsworth Publishing Co., Belmont, California.

Šestak, Z., J. Čatský and P.G. Jarvis, editors. 1971. Plant Photosynthetic Production, Manual of Methods. Junk, The Hague, Netherlands.

Slatyer, R.O. 1967. Plant-Water Relationships. Academic Press, New York, NY.

Spomer, L.A. 1980. Guidelines for measuring and reporting environmental factors in controlled environment facilities. Commun. Soil Science and Plant Analysis 11(12):1203-1208.

Spomer, L.A. 1981. Guidelines for measuring and reporting environmental factors in growth chambers. Agronomy Journal 73(2):376-378.

Thimijan, R.W. and R.D. Heins. 1983. Photometric, radiometric, and quantum light units of measure: A review of procedures for interconversion. HortScience 18(6):818-821.

Tibbitts, T.W. and T.T. Kozlowski, editors. 1979. Controlled Environment Guidelines for Plant Research. Academic Press, New York, NY.

Tooming, K.G. 1977. Solar Radiation and Yield Formation (Solnechnaya radiatsiya i formirovanio urozhaya) Gidrometeoizdat, Leningrad.

Zelitch, I. 1971. Photosynthesis, Photorespiration and Plant Productivity. Academic Press, New York, NY.

INDEX